Telecommuting

For a complete listing of the *Artech House Telecommunications Library*,
turn to the back of this book.

Telecommuting

Osman E. Eldib
Bell Communications Research, Inc.

Daniel Minoli
DVI Communications
Stevens Institute of Technology

Artech House
Boston • London

Library of Congress Cataloging-in-Publication Data
Eldib, Osman.
Telecommuting / Osman Eldib and Daniel Minoli
Includes bibliographical references and index.
ISBN 0-89006-738-4 (acid-free paper)
1. Telecommuting—United States. 2. Telecommunication—United States.
I. Minoli, Daniel, 1952– . II. Title.
HD2336.U5E43 1995 94-49213
331.25—dc20 CIP

A catalogue record for this book is available from the British Library

Eldib, Osman
Telecommuting
I. Title II. Minoli, Daniel
621.382

ISBN 0-89006-738-4

685 Canton Street
Norwood, MA 02062

International Standard Book Number: 0-89006-738-4
Library of Congress Catalog Card Number: 94-49213

10 9 8 7 6 5 4 3 2 1

For my wife, Audrey, and my parents
(O.E.E.)

For Gino, Angela, Anna, Emmanuelle, Emile, and Gabrielle
(D.M.)

Contents

Preface

The growing interest in telecommuting has been reflected in a wide range of reports, articles, newsletters, telephone company trials, and products and services focusing on telecommuting. This interest has been driven by a growing telecommuting industry, which is attracting an increasing number of people from the following areas:

- *Employers from a variety of industries*, such as education, government, and manufacturing, are currently either implementing telecommuting programs or considering the implementation of such programs.
- *Employees with a diverse range of demographic characteristics* (e.g., age, occupation, geographic locations) are seeking telecommuting as an alternative to commuting to work.
- *Network service providers*, such as interexchange carriers (IXCs), local exchange carriers (LECs), alternate access providers (AAPs), and cable television companies are increasingly competing with each other and are exploring new business opportunities, including those in the telecommuting market.
- *Providers of applications solutions*, such as value-added network (VAN) providers, Internet providers, and groupware vendors, are also exploring new business opportunities.

This book examines the topic of telecommuting from both technical and business perspectives. The book is targeted at the corporate employers involved in the planning, development, and implementation of telecommuting programs as well as at existing and potential telecommuters. The goal of the book is to provide the corporate decision makers and telecommuters with a *guide to telecommuting communications applications and networking solutions*. To accomplish this goal, several objectives are established, as follows.

1. The key existing and potential buyers of telecommuting services are identified and the factors driving their needs for telecommuting are outlined.
2. The key groups of corporate decision makers that participate in the planning, development, and implementation of telecommuting programs are identified and the role of each group in the decision making process is described.
3. The most likely telecommuters are characterized and the functions they perform are defined.
4. The key communications applications of telecommuters are defined and communications requirements of telecommuters associated with these applications are highlighted.
5. A number of major networking solutions that address telecommuters' applications requirements are described and the strengths and shortcomings of these solutions are compared.
6. Current and emerging providers of these networking solutions are introduced, and the key communications services that they offer are explored.
7. Specific actions that buyers, communications, and applications providers need to take to maximize their chances of success in the telecommuting industry are outlined.

Current corporate employers and employees that are considering, implementing, or participating in, a telecommuting program can benefit from this book. Employers can learn the *communications technology* factors as well as the administrative and human factors that they should take into account in creating a successful telecommuting program (or enhancing an existing program), which aims at increasing the productivity of their employees and reducing operating cost. Employees can become more familiar with the technical factors associated with working at home, enabling them to develop and present more effective business cases to their employers in support of telecommuting. Information systems (IS) managers can learn about the existing and emerging networking solutions that are (or will be) available to them, along with the strength and limitations of each solution.

The book could also be beneficial to enhanced service providers and network service providers that have not entered the telecommuting market yet and wish to learn about the needs of the market and to what extent their current telecommunications infrastructure and services meet these needs.

Acknowledgments

Mr. Eldib would like to thank Mr. John Mulligan and Mr. Don Tow of Bellcore for their valuable contributions to this text. Mr. Eldib would also like to thank Mr. Frank Gratzer, Bellcore, for his support.

Mr. Minoli would like to thank DVI Communications and Bellcore for their moral support of this work. Mr. Minoli would also like to thank Mr. Ben Occhiogrosso, DVI Communications, for making available information about telecommuting analysis methodologies used by the company in support of its consulting practice.

Note

This book does not reflect any policy, position, or posture of any corporation or institution. All ideas expressed are those of the authors. The writing of the book was not funded by anyone. Data pertaining to the public switched network is based on the open literature, and was not reviewed by respective carriers. Vendor products, services, and equipment, are mentioned solely to document the state of the art of a given technology, and the information has not been counter-verified with vendors. No material contained in this book should be construed as a recommendation of any kind.

Osman E. Eldib
Daniel Minoli

An Overview of the Telecommuting Industry 1

In many large and/or fast-growing U.S. cities, the commuting chore is becoming more demanding. A 1994 study by the Federal Highway Administration showed that getting to and from work is taking longer now than a decade ago as jobs and homes move to the suburbs, away from arterial highways. In 1980 the average commute was 21.7 minutes; in 1990 the average commute was 22.4 minutes. Therefore, an increasing number of corporations and employees are looking at alternatives. These alternatives fall in the area of what has been called "telecommuting." This chapter explains the concept of telecommuting and introduces the reader to the key elements of the telecommuting industry structure that are further examined throughout the book. These elements include the key buyers, the key networking and application solutions that address the requirements of these buyers, and the different types of suppliers that offer these solutions. This chapter is concluded by describing the structure and logical sequence of the remaining chapters in this book.

1.1 DEFINITION OF TELECOMMUTING

Telecommuting is a term coined in the recent past to refer to the ability of workers to either work out of their homes or to only drive a few minutes and reach a complex in their immediate neighborhood where, through advanced communication and computing support provided by the "landlord" of the complex, they can access their corporate computing resources and undertake work. Eventually there could be at least one such complex, known as "telebusiness center" [1] or regional telecommuting center (RTC), in each suburban town. In addition to data and voice capabilities provided on an economy-of-scale basis, these facilities could include shared or nonshared videoconferencing systems, PCs, workstations, scanners, color printers, photocopiers, and other office equipment. In the shorter term, however, telecommuting is likely to be home-

or home office-based. One of the goals of the designers of a telecommuting program is to duplicate the communications environment that employees have in their offices at their home [2].

Despite official encouragement in favor of carpooling, a 1994 study by the Federal Highway Administration found that more people now commute alone—the number increased between 35% and 73% in one decade (in reference to any kind of driving averaged over the United States, 2.4% more people were driving to work in 1990 compared to 1980.) The study also showed that in 1990, 73.2% of the United States workers drove alone to work, 13.4% carpooled, 5.3% used mass transit, 3.9% walked, and 3% worked at home (1.2% used other means such as boats, planes, bicycles, etc.). Such driving habits increase pollution and congestion. In the past, few alternatives were available; now telecommuting technologies and services can provide a relief from the burdens of physical commutation to work.

In the next 18 months, telecommuting is going to be catapulted center stage by regulatory, technological, productivity, and practical forces. Organizations need to start to make appropriate plans at this time in order to position themselves in a proactive manner and not be caught lagging behind implementation life cycles, unprepared to meet the mandated timetables for conformance. Cities such as New York, Houston, and Los Angeles are specifically targeted as early sites of conformance. Affected organizations include corporations, real estate concerns, and public works/public planning agencies of municipalities, counties, and states. Within corporations, management, facility planning, and information technology groups will be impacted.

Telecommuting can also be seen in the context of the evolution of the mobile professional: such professionals may actually work at the office for eight hours and/or a few days a week and wish to have a "replica" of their office at home, on the road, or even in the "random" hotel room they may find themselves in during a business trip [3,4]. Some companies make ample use of wireless technology both in the office (for both voice and data) as well as in the field. This includes portable computers, cellular phones, pagers, text pagers, personal digital assistants (PDA), and other mobile computing and communications devices. At press time there were already seven million people that do at least part of their work on the road [5]. The telecommunication and supporting technologies that support a home office can often (but not always) be used to support the mobile worker.

A variant of telecommuting is what has been called the *virtual office*, also known as *hoteling*, a concept that has emerged in the recent past [5–8]. In short, hoteling implies asking workers to work outside the office, visiting clients, working at home, or otherwise being in the field, and reducing the office space so that at any one time only 30% to 50% of the workforce could be present and seated in the office. No permanent offices are assigned, and all offices are stark and identical; when a worker needs to come to the office to accomplish a task that cannot be done on the road, or for some meeting, the worker is assigned a cubicle from a pool of available seating positions, much the same as a person is assigned a room when he or she goes to a hotel (one might call that a "cubicle de jour"). Each cubicle has voice and data jacks so that workers can connect their laptops or PDAs, and be connected with the company file, application, and e-mail server. A system, typically new software that is loaded into the office PBX, assigns the available cubicle to the user and then routes all incoming calls for that particular user to the appropriate cubicle.

The initial applications of hoteling were in the context of company re-engineering, in conjunction with downsizing, business process redesign, or office space consolidation and saving. Companies that have implemented the concept, and which are reportedly committed to expending its deployment include IBM, Ernst & Young, Arthur Anderson, Milliken, and Chiat/Day [5–7]. Ernst & Young reduced its space needs by about 20% (450 accountants and consultants share 100 seating positions), while IBM reduced its needs by 75% in the offices where hoteling has been used [7]. One can envision a situation where these techniques are put in place to support occupancy management at a regional telecommuting center. Again, the PBX or equivalent software appropriately assigns the cubicle and supporting communications.

Hoteling techniques can free knowledge workers, managers, planners, and executives from the office. However, there is no current consensus on the overall employee and employer acceptance and its popularity in the future. Some see this as the "wave of the future" as jobs for life are vanishing and the laptop computer becomes an icon of traveling office.

There have been human-factor challenges to this nonterritorial office concept. However, some of the companies that have tried it allow their employees to "work from any location they choose;" these proponents believe that they cannot "institutionalize things just to be able to conveniently manage them" [5]. Employee-to-employee interactions are supported via e-mail, file sharing, and other innovative workgroup software (e.g., graphically driven groupware interfaces that realistically reflect the virtual elements of the virtual office).

All seven of IBM's regional U.S. market offices were reportedly contemplating the virtual office approach. The first office converted was in New Jersey, where 400,000 square feet of prime space was vacated in favor of a high-vaulted steel-walled industrial building without enclosed offices, which instead contained 220 shared desks for 600 employees [6,7].

From a business standpoint, telecommuting can be viewed as a segment of the work-at-home market (Figure 1.1). It encompasses those employees of corporations, government agencies and universities who work at home either part time or full time during business hours for at least eight hours per week. The work-at-home market encompasses other segments beyond the telecommuting market, including full-time self-employed workers, part-time self-employed workers, and corporate after-hour workers (these segments are not addressed in this book).

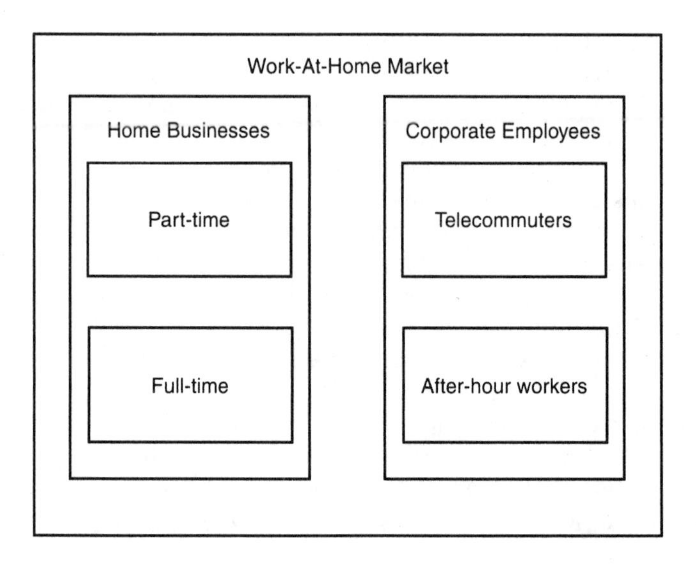

Figure 1.1 The work-at-home market segments.

The U.S. areas with the most congestion in 1993, where telecommuting makes immediate sense, were:

- Los Angeles;
- Washington;
- San Francisco-Oakland;
- Miami;
- Chicago;
- San Diego;
- Seattle-Everett;
- San Bernardino;
- New York;
- Houston.

Tables 1.1–3[1] provides some parametric information related to commuting; information such as this may be factored in by the designers of a telecommuting program. Note that at a nominal $20 per hour, a 60-minute round-trip commute equates to $5,000 per year in productivity loss per employee in addition to the actual costs of transportation and the "hidden" environmental costs. Knowledge workers are typically paid more and so telecommuting makes a lot of sense from a productivity point of view.

1. These tables are based on data from the Federal Highway Administration.

Table 1.1
Recent Statistics about Commuting Trends

City where the largest number of people drive alone	Detroit (82.7%)
City where the least number of people drive alone	New York City (52.3%)
City where the largest number of people carpool	Washington DC (15.8%)
City where the least number of people carpool	Boston (10.3%)
City where the largest number of people use mass transit	New York City (27.8%)
City where the least number of people use mass transit	Tampa (1.6%)

Table 1.2
Average Commutation Time by City (1990)

City	*Time (minutes)*
Buffalo	19.4
Providence	19.6
Rochester	19.7
Salt Lake City	19.8
Milwaukee	20.0
Hartford	20.6
Minneapolis	21.1
Columbus	21.2
Kansas City	21.4
Charlotte	21.6
Norfolk	21.6
Portland	21.7
Sacramento	21.8
Tampa	21.8
San Antonio	21.9
Indianapolis	21.9
Cleveland	22.0

Table 1.2 (continued)
Average Commutation Time by City (1990)

City	*Time (minutes)*
Cincinnati	22.1
San Diego	22.2
Denver	22.4
Pittsburgh	22.6
Orlando	22.9
Phoenix	23.0
St. Louis	23.1
Detroit	23.4
Dallas	24.1
Miami	24.1
Philadelphia	24.1
Boston	24.2
Seattle	24.3
New Orleans	24.4
San Francisco	25.6
Atlanta	26.0
Baltimore	26.0
Los Angeles	26.4
Chicago	28.1
Washington	29.5
Houston	29.5
New York	31.1

As an industry, telecommuting has evolved in the last several years to encompass an increasing number of buyers, communications equipment, substitute and complementary telecommuting services, applications solutions, and network and application solutions providers. These elements of the telecommuting industry structure are introduced in this chapter.

Table 1.3
Changes in Commutation Time by City (1980–1990)

Cities With Highest Change in Commutation Time 1980-1990	*Percent Change Between 1980 and 1990*
San Diego	13.7
Orlando	12.7
Los Angeles	11.9
Sacramento	11.8
Charlotte	8.6

1.2 BUYERS OF TELECOMMUTING SERVICES

Current and potential buyers of telecommuting services include employers in: government agencies at the state, local, and federal level; universities, colleges, and other higher education institutions; and a growing number of corporations. This market also includes employees in these industries who perform a wide range of technical and administrative functions. The functions that telecommuters perform and the number of telecommuters determine the networking access requirements and the network aggregation requirements at the corporate end of the telecommuting system.

The telecommuting market is currently underdeveloped. It can be synthesized on a top-down as well as bottom-up basis. Traditionally one looks at the telecommuting market from the perspective of the small business/work-at-home segment. This is a bottom-up approach. This segment is cash-poor, and is not necessarily the group to which a provider can directly sell services (with some exceptions). In this text, telecommuters are taxonomized for convenience into administrative, technical, marketing, and home-agent telecommuters.

A top-down view of the market, where a provider can directly sell services, is as follows.

1. *Corporate planners and strategists.* These are individuals in corporations that will need to become involved in meeting the implications of the Clean Air Act and related legislation. The issue is: who in the corporation makes such decision? Senior management? The office of the president? The CIO? One may identify who to contact by considering companies headquartered in the geographic location of interest. One can also focus on specific industries perhaps more open to the concept (for example, banks may not be willing because of security considerations, but publishing firms could be interested). One can also focus on

specific functions within corporations (for example, some functions may be more adequate to telecommuting than others).

2. *Real estate concerns.* Real estate developers may worry about a decreased demand on office space. However, looking at telecommuting as an opportunity, real estate developers could establish regional telecommuting centers either by building new facilities strategically located or by equipping an existing building with the necessary communications equipment.
3. *Regional (state, county, city) planners.* Infrastructure planners (highways, rail, bridges, etc.) may become involved in the process of looking for a solution to a congestion problem. For example, in studying the alternatives of either building a new highway system or developing a "teleport" hub including a regional telecommuting center, they may find that the latter is much more cost effective.
4. *New services developers.* Carriers, new service developers, and equipment manufacturers may specifically target telecommuting services. In particular, they may need assistance from consultants that know the field, the technology, and the end-user in order to roll out communication services.
5. *Entrepreneurs.* Hotels, Kinkos, and so forth may be looking at plans to develop an overall telecommuting posture for entry into the telecommuting-support business.
6. *Local universities.* Local universities may wish to focus on a special type of telecommuting: distance learning.
7. *Market researchers.* Market researchers may find that the information they are able to generate on telecommuting is of interest to the parties identified above (government agencies, corporate planners, carriers, etc.).

1.3 COMPUTING AND COMMUNICATIONS EQUIPMENT

To work at home, telecommuters need several types of communications and computing equipment. Some of the basic equipment are introduced next.

1.3.1 Personal Computers

Personal computers (PCs) are increasingly being used by telecommuters. According an industry forecast, in 1992, 21.3% of telecommuting households (up from 15.2% in 1991) used a PC to work at home [9]. The growth in the use of PCs can be attributed to several reasons: (1) PC literacy is on the rise as schools produce students that are familiar with PCs and their applications, (2) computer manufacturers have been introducing computers with friendlier interfaces; examples of these computers include Apple Computer Inc.'s System 7.0 for the

Macintosh and Microsoft's Windows for both IBM and *clone* PCs, and (3) the prices of computers have been dropping as a result of the price wars among computer manufacturers. A telecommuter can now purchase a reasonably-configured PC at a price as low as $1,000 [10].

1.3.2 Telephone Sets

Telecommuters may rely on basic telephone sets, cordless phones, or cellular phones. Basic telephone sets remain the most predominant communications tool of telecommuters. They were used by over 90% of those working at home in 1992. Cordless phones are used by almost half the work-at-home households [11], while cellular phones are at an earlier stage of development (these are used more by mobile/nomadic telecommuters): only 5.6% of those working at home used cellular phones. According to a recent industry forecast, only 10.4% of those working at home were interested in purchasing cellular phones, while 30.4% are somewhat interested in cellular phones. The degree of interest in owning a cellular phone and the type of cellular phone used by a telecommuter depend upon the function performed by the telecommuter and his or her degree of mobility. (The specific mobile communications needs of the telecommuters are explored later in Chapter 5.)

1.3.3 Fax Machines

As in the case of cellular phones, fax machines have not been widely purchased by those working at home. According to industry forecasts, only 4% of those working at home used fax machines in 1992. The demand for fax machines is likely to rise. The degree of interest in owning a fax machine, and the type of fax machine used by a telecommuter depend upon the function performed by the telecommuter and his or her degree of mobility.

1.3.4 Copying Machines

In 1992, the number of telecommuters using copying machines was slightly higher than those using fax machines. According to industry forecasts, 7.8% of those working at home used copiers in 1992, slightly more than the number of those working at home that used copying machines in 1991 (6.5%). Purchasing levels of copying machines are not likely to increase because telecommuters will increasingly purchase fax machines to perform faxing as well as copying functions.

1.3.5 Modems

Modems are key elements of the dialup solution, introduced in Section 1.4.1.

1.4 INTRODUCTION TO NETWORK SOLUTIONS

Clearly, communication services play a central role in telecommuting. Several network solutions exist to support the communications requirements of the telecommuters (Figure 1.2). These solutions can be classified into three major groups:

- *Narrowband* solutions, which operate at speeds up to 1.5 Mbps.
- *Wideband* solutions, which operate at speed above 1.5 Mbps but below 45 Mbps.
- *Broadband* solutions, which operate at speeds of 45 Mbps and above.

While narrowband solutions are appropriate for individual telecommuters working out of their homes, wideband and broadband solutions are more appropriate for regional telecommuting centers and some corporate offices. This text explores a wide range of narrowband, wideband, and broadband solutions. These solutions are introduced next.

1.4.1 Dialup

This plain old telephone service (POTS) approach involves the use of modems to connect the telecommuter PC (owned by the telecommuter or provided by the corporation) to the remcte server or mainframe; it utilizes the analog public telephone network. Circuit switching implies that the communications channel is not dedicated 24 hours per day, but must be brought online when needed (via a process called *call setup*) and then taken down when no longer needed. Traditional modems have operated at speeds up to 28.8 Kbps; however, until recently, 9.6 Kbps has been more common (see Table 1.4). This implies that the throughput across this type of link is fairly small; consequently, only a small number of users and/or short inquiry/response-like transactions can be supported. Since the link between the two points is not available on a dedicated basis, the user needs to dial up the corporate data center or access node, as needed.

1.4.2 Data Over Voice

Data over voice (DOV) refers to a group of existing technologies that are used to transmit data and voice over the same copper pair. Data and voice are sent in separate channels that operate in different frequency ranges. DOV currently exists in three forms: (1) analog voice and data, (2) analog voice, digital data, and (3) digital voice and digital data. DOV-based services are currently offered on a limited basis on some U.S. regions.

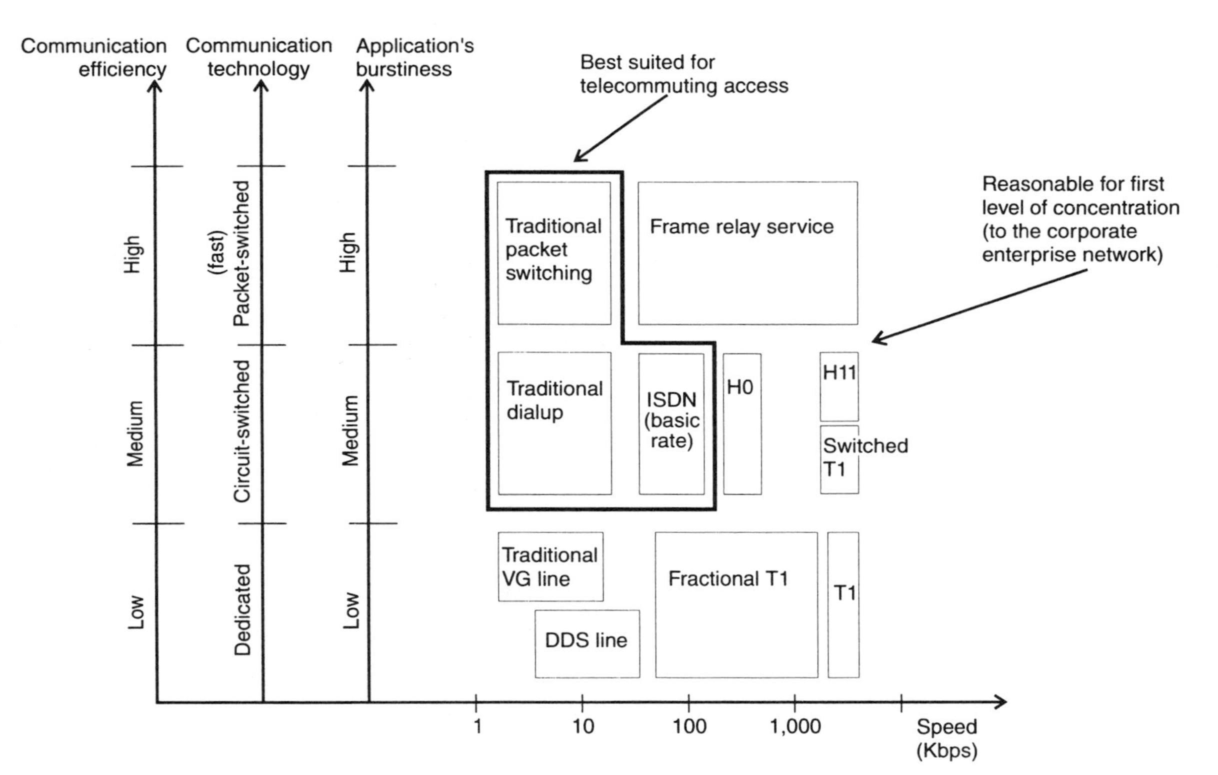

Figure 1.2 Telecommunication services that can support telecommuting.

Table 1.4
The Evolution of the Dialup Modem

Standard	*Highest Data Rate (bps)*	*Introduced*
Bell 103	300	1960s
Bell 212 (v.22)	1,200	1978
V.22*bis*	2,400	1981
V.32	9,600	1986
V.32*bis*	14,400	1991
V.*fast (v.34)*	28,800	1994

Table partially based on [12].

1.4.3 Packet Switching

Packet switching, a technology that first arose in the mid-1960s, affords statistical allocation of bandwidth. Information is exchanged as blocks of limited size or *packets*. At the source, files (and other types of data blocks) are partitioned into an appropriate number of packets; packets are transmitted across the network and are reassembled at the destination to reconstitute the original file (or data block). Multiple users can share network resources, thereby lowering the costs. Packet-switching service can be obtained via a privately owned network or via a public packet-switched carrier. Packet switching has been standardized according to the ITU-T (International Telecommunications Union-Telecommunications, formerly, CCITT) recommendation X.25, first published in 1976 (the recommendation was significantly revised in both 1980 and 1984; minor revisions have taken place since). Packet switching's throughput has traditionally been limited to around 9.6 Kbps (more recently, 56 Kbps, particularly in the backbone) and, hence, is not ideally positioned to support all of today's data-intensive business applications, although some users could still employ it, for example for e-mail, store-and-forward, and information access applications (e.g., CompuServe). Packet-based networks are typically priced on a timed-usage basis, and are therefore sensitive to data volume; they are, however, insensitive to distance and are only slightly sensitive to the number of sites added. This service is typically provided by value-added networks (VANs) at the national level and by some LECs at the local level.

1.4.4 Integrated Services Digital Network

This interconnection approach involves the use of switched *digital* facilities between the telecommuter and the corporate access node or data center. ISDN provides end-to-end digital connectivity with access to voice and data (packet and circuit) services over the same digital transmission and switching facilities. It provides a range of services using a limited set of connection types and multipurpose user-network interface arrangements. ISDN supports narrowband solutions through three channels types: B channels, D channels, and H channels. The *B channel* is a 64-Kbps access channel that carries customer information such as voice calls, circuit-switched data, or packet-switched data. The *D channel* is an access channel carrying control or signaling information and, optionally, packetized customer information; the D channel has a capacity of 16 Kbps or 64 Kbps. The *H channel* is a 384-Kbps, 1.536-Mbps, or 1.920-Mbps (Europe) channel that carries user information such as video teleconferencing, high-speed data, high-quality audio or sound programs, and imaging data.

ISDN defines *physical user-network interfaces*. The more well-known of these interfaces is 2B + D, which provides two switched 64-Kbps channels plus a 16-Kbps packet/signaling channel (144 Kbps total); other well-known interfaces are shown in Table 1.5. Other less-known ISDN interfaces include 1B + D, which are currently trialed by some LECs (e.g., NYNEX), and NB + D where N is greater than 2 but less than 23.

Table 1.5
ISDN Interfaces

23B + D	23 switched 64-Kbps channels plus a 64-Kbps packet/signaling channel (1.536 Mbps total)
H0 + D	Switched aggregated 384-Kbps links
H10 + D	Switched aggregated 1.544-Mbps links.

According to some observers, no single ISDN market is garnering as much enthusiasm as the telecommuting market; these proponents see "solid potential for ISDN to fill telecommuting needs" [12–15]. However, this assessment on the potential of ISDN may be somewhat optimistic. Many users now employ faster modems, operating at 14.4, 19.2, and 28.8 Kbps without compression, and 57.6, 115.2, and 230 Kbps with compression. In favor of ISDN is the fact that it offers a more "comfortable" 64-, 128-, or even 384-Kbps throughput, affording adequate flexibility to the user. A number of carriers are also introducing a single line ISDN for customers (such as telecommuters) that have neither a PBX or Centrex service [4]. Some see a combination of ISDN and frame relay as a viable way to bring in remote users and connect them to the enterprise network (which in this case would be based on frame relay services).

1.4.5 Digital Private Lines (Fractional T1, T1, T3)

Dedicated private line digital services provide transparent bandwidth at the specified speed ($N \times 64$ Kbps for FT1, 1.544 Mbps for T1, and 44.736 Mbps for T3) and are suitable for point-to-point interconnection of low-burstiness (i.e., steady) traffic. T3 (also known as DS3) facilities are increasingly available in many parts of the country, although they are still fairly expensive. In addition, a number of carriers have started to offer a fractional T3 service that allows the user to specify the desired number multiple of T1s. These services may be more appropriate for RTCs than for individual telecommuters. Figure 1.3 depicts a traditional enterprise network that uses private lines for the backbone and ISDN for access [16].

1.4.6 Frame Relay Service

Frame relay service is a data communication service that became available in the early 1990s. It supports medium-speed connections between user equipment (routers and private switches in particular), and between user equipment and carriers' frame relay network equipment (i.e., public switches). The frame relay protocol supports data transmission over a connection-oriented path; it enables the transmission of variable-length data units, typically up to 4,096 octets over an assigned virtual connection. Frame relay service provides interconnection among n (backbone) sites by requiring only that each site be connected to the "network cloud" via an access line (compare this with the $n(n-1)/2$ end-to-end lines required with dedicated services). The cloud consists of switching nodes interconnected by trunks used to carry traffic aggregated from many users. In a public frame relay network, the switches and the trunks are put in place by a carrier for use by many corporations. Carrier networks based on frame relay provide communications at up to 1.544 Mbps (in the United States), shared bandwidth on demand, and multiple user sessions over a single access line. Compared to packet switching, the frame relay service provides more throughput, which is attributed to higher bandwidth and lower protocol overhead. The higher throughput makes the service attractive for telecommuting applications that involve image transfer. In a private frame relay network, the switches and trunks are put in place (typically) by the corporate communications department of the company in question. In either case, the service can provide permanent virtual connections (PVCs).

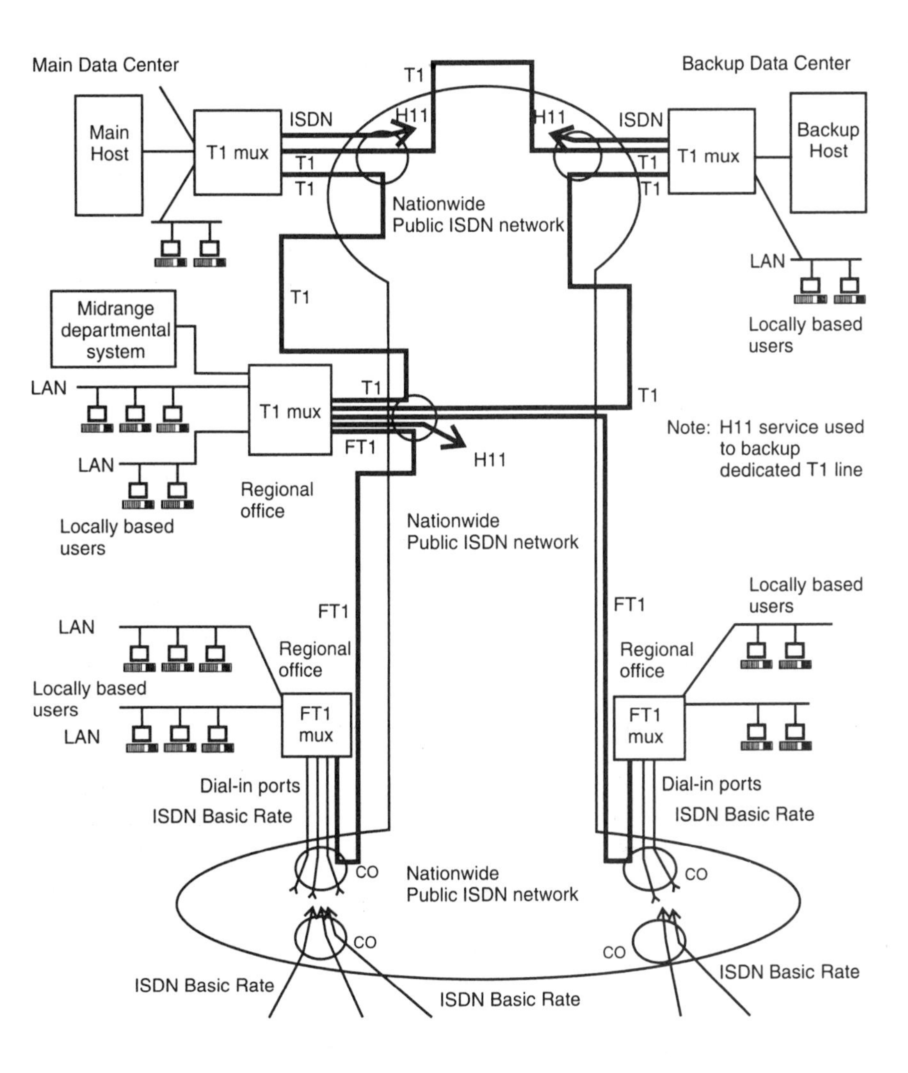

Figure 1.3 An example of a traditional enterprise network.

1.4.7 Switched Multimegabit Data Service

Switched multimegabit data service (SMDS) is a high-performance, public, connectionless datagram service that was developed by Bell Communications Research (Bellcore). Access is currently based on a subset of the IEEE 802.6 metropolitan area network (MAN) standards. Access to SMDS is established through a subscriber network interface (SNI), which may provide either a DS3 access path (with multiple access classes) or DS1 access path. For the DS3 access path, single or multiple-CPE access arrangements are available. SMDS interface protocol (SIP) is based on IEEE 802.6 distributed queue dual bus (DQDB) connectionless protocol. This protocol operates at the MAC level of the data link layer (layer 2 of the OSI reference model). As a result, SMDS can be supported by various higher layer protocols, such as TCP/IP and DECNET.

SMDS supports five access classes. One of these access speeds can be supported by a DS1 access line, while the other speeds (4 Mbps, 10 Mbps, 16 Mbps, 25 Mbps, and 34 Mbps) can be obtained by using a portion of a DS3 line. SMDS may also operate in the future at 64 Kbps.

1.4.8 ATM/Cell Relay Service

Asynchronous transfer mode (ATM) refers to a high-bandwidth, low-delay switching and multiplexing technology now becoming available for both public and private networks. While ATM in the strict sense is simply a (data link layer) protocol, the more encompassing ATM principles and ATM-based platforms form the foundation for a variety of high-speed digital communication services aimed at corporate users for high-speed data, LAN interconnection, imaging, and multimedia applications. Cell relay service is one of the key services enabled by ATM. Cell relay service [17] can be utilized for enterprise networks that use completely private communication facilities, completely-public communication facilities, or that are hybrid. A variety of vendors are now readying end-user products for market introduction; some products were already been on the market a couple of years ago. A number of carriers either already provide services or are poised to do so in the immediate future. ATM supports both switched (SVC) and nonswitched (PVC) connections. ATM supports services requiring both circuit-mode and packet-mode information transfer capabilities. ATM can be used to support both connection-oriented and connectionless services (e.g., SMDS).

1.4.9 Native-Mode LAN Interconnection Service

Native-mode LAN interconnection service (NMLIS) is a family of LAN extension and channel extension services designed to support connectivity between the same types of LANs (Ethernet, token ring, and fiber distributed data interface) and high-performance computer systems across MANs and wide area networks (WANs). These services operate at native LAN speeds (4, 10, 16, 100 Mbps) and can support the LAN interconnection requirements of RTCs, providing RTCs with a substitute to the following services: frame relay, SMDS, and private T1 services. The underlying technology can, however, be any of these, in addition to dedicated lines and/or dark fiber.

1.4.10 Integrated Access

Integrated access is an emerging service that complements frame relay, SMDS, private lines, and voice services by enabling the RTC to gain access to these services and to other dedicated services through a single access line. The integrated access line aggregates various types of customer traffic (e.g., PBX, video). Aggregated traffic is transmitted from the customer premises to a telephone company central office (CO) where it is deaggregated and routed to the appropriate WAN services (e.g., FRS, PSTN, and X.25).

1.4.11 Asymmetric digital subscriber line

Asymmetric digital subscriber line (ADSL) is an emerging high-performance copper-based technology currently being standardized by the T1E1.4 subcommittee of the Alliance for Telecommunications Solutions. ADSL has extensive capabilities, including one bidirectional 160-Kbps ISDN basic rate channel containing two B channels, a D channel, and associated ISDN operations channels; a control channel; an embedded operations channel for internal systems maintenance, audits, and surveillance (not directly visible to the user); and passive coupling of ADSL to basic phone service. ADSL is currently being evaluated or trialed by the RBOCs, such as Bell Atlantic.

Substitute Versus Complementary Networking Solutions

The narrowband and wideband solutions described above provide substitutes to each other, each with a set of strengths and weaknesses. Some of these solutions can also complement each other, providing the users with hybrid solutions. Examples of complementary networking solutions include frame relay/ISDN, Cable TV/POTS, and ATM/SMDS.

1.5 APPLICATIONS SOLUTIONS

The networking solutions described above operate at lower layers of the open systems interconnection reference model (OSIRM), enabling telecommuters to establish communications links with each other as well as with trading partners (e.g., suppliers and customers.) In addition to these networking solutions, other *application-level solutions* are explored in this book, including groupware, the Internet, and VANs. The following is a brief introduction to each applications solution.

1.5.1 Groupware

Groupware refers to software that may support at least one of the following applications:

- Electronic messaging;
- Data conferencing;
- Process management;
- Messaging gateways.

An increasing number of vendors offer groupware. The products offered by these vendors can be classified into three categories:

1. *Low-end groupware* that supports a single application, such as computer conferencing, project management, workflow management, and time management;
2. *Midrange groupware* that supports a single local area or workgroup, but may support a number of functions;
3. *High-end software* that supports multiple functions over wide area enterprise networks.

1.5.2 The Internet

The Internet is an international network that supports over 34,000 computer networks and over 2 million computers worldwide. Estimates of the number of Internet users vary from 7 million to 12 million; half the Internet users reside in the United States. The Internet has been growing rapidly in the last several years. Several statistics reflect this fact: the number of subscribers has been growing at 100% per year for the last three years; the number of universities and research laboratories using the Internet has increased from a few hundred in 1988 to more than 10,000 in 1993; and, the number of packets carried on the Internet has increased from millions per day to billions per day. Through the Internet, users can receive a number of services, including logon services, e-mail, file transfer, host-to-host communications and directory services.

1.5.3 Value-Added Networks

Value-added networks (VAN) provide telecommuters who need to communicate with their customers, suppliers, and corporate offices with an alternative to the Internet. VANs are X.25-based networks and provide subscribers with a number of services, including e-mail, X.400, database retrieval, integrated messaging, SNA support, closed user groups, and other enhanced services.

While VAN and Internet solutions started as competitive alternatives, an increasing number of VAN providers are providing their subscribers with access to the Internet.

1.6 NETWORK AND APPLICATIONS SOLUTION PROVIDERS

Networking and applications solutions introduced above are offered by an increasing number of *network service providers* including local exchange carriers (LECs), interexchange carriers (IXCs) and cable TV providers. The range of services offered differs from one group of providers to another. The range also differs from one provider to another. Some of these services are targeted specifically at the telecommuting market.

In addition to the existing network service providers, several *new entrants* are likely to enter the telecommuting market in the next few years, intensifying the competition in this market. New entrants may include (1) cable TV companies, (2) independent LECs, and (3) additional IXCs. These existing and emerging network service providers are being supported by *hardware and software providers* that have recognized the importance of the telecommuting industry as a potential source of revenues. These companies either sell directly to the telecommuting buyers or support the network service providers in providing these buyers with telecommuting-related services. The supply structure of the telecommuting industry also includes a number of providers of applications solutions, including VAN service providers, Internet access providers, and groupware vendors.

1.7 ANALYTICAL METHODOLOGIES IN SUPPORT OF TELECOMMUTING

Telecommuting is a fairly complex multidimensional, multidisciplinary problem at the strategic decision-making level. As noted above, there are multiple constituencies, multiple network technologies, and a variety of user equipment. A methodical approach is needed by decision makers to sort through numerous alternatives, options, and approaches. As an example, DVI Communications, New York, NY, has developed a systematic methodology to assist organizations in developing and deploying an effective telecommuting program. DVI's TELEMETHOD (telecommuting methodology for optimized decision-making), like the subject itself, spans the multiple disciplines that need to be taken under consideration. These disciplines include information technology, communications, security, oversight, human factors, productivity, regulatory, and compensation. In the past, a number of organizations have started pilot programs with mixed success; at this juncture, organizations cannot afford the risk of false starts. A sophisticated methodology enables organizations to immediately focus on the required sequence of planning tasks and decision points.

Regulation to be taken into account includes The Clean Air Act Compliance, the National Emergency Legislation, the Internodal Surface Transportation Efficiency, and the American Disabilities Act. The methodology incorporated in TELEMETHOD overlays these regulatory requirements onto the industry segment of which the target corporation is a member, the corporation's workflow structure, and its specific mode of operation. The intersection of the legislative requirements with the corporate workflow requirements leads to a set of primitive bases, which are then used to support the analysis portion of the methodology. The analysis results in decision matrixes to be used by the organization planners, related to information processing, communications, security, human factors, and oversight.

A sophisticated methodology enables the planner to answer the following questions, among others.

1. What is the impact of the telecommuting regulation on the organization?
2. Does the regulation affect an industry more than another?
3. When must an organization start to put in place a plan or system?
4. How extensive must such a plan/system be?
5. What impact will telecommuting have of the productivity of the organization?
6. Which functions are better suited to telecommuting and which are not?
7. What impact will telecommuting have on the managerial and oversight function?
8. What are the real estate/office space implications?
9. What are the legal and insurance implications of telecommuting?
10. What are the data processing implications? Which data processing architectures are more amenable to the support of telecommuters?
11. What are the telecommunication implications? Will the cost of doing business go up because of the increased reliance on communication services? How does one have to modify the organization's enterprise network to support telecommuting? How does one support corporate voice-transmission requirements? Data transmission requirements? Video conferencing requirements?
12. Are there technologies that are particularly suited to support telecommuting? Is ISDN the right answer? Wireless? Cable systems and/or RBOCs' video dialtone systems? Commercial Internet providers? Will connect charges become a major expense item?
13. How can telecommuting and outsourcing be combined to achieve maximum savings for organizations?

This list of questions reflects the complexity of the issue, which goes well beyond mere technical considerations.

1.8 COURSE OF INVESTIGATION

The preceding discussion presented an overview of major telecommuting industry forces—the buyers, the basic and complementary applications and networking solutions, the existing network and applications service providers, the new entrants, and the equipment suppliers. The themes introduced in the previous sections will be explored in greater detail in the course of this text. The treatment is divided into ten chapters.

Chapter 2 provides a macroanalysis of the telecommuting demand, including demand side, growth, and distribution. Demand distribution is examined along several dimensions, including geography, company size, and vertical industry. This chapter then identifies telecommuting demand drivers and demand inhibitors. The chapter is concluded by identifying the key corporate players involved in telecommuting, including (1) the decision makers at the corporate level who are involved in the planning, development, and implementation of a corporate telecommuting program; and (2) the key groups of telecommuters.

Chapters 3 through 9 provide a detailed description of a number of telecommuting communications applications, including the following.

- Basic voice and data communications applications;
- High bandwidth data applications;
- Mobile data applications;
- Video communications applications;
- Computer telephone integration applications;
- Regional telecommuting center applications.

Chapters 3 through 9 link these applications to specific groups of telecommuters and explains how these applications can enhance the value that the telecommuters create to their employers. These chapters also explore a number of narrowband and wideband solutions to meet the requirements of these applications. The following is a brief description of the issues addressed in Chapters 3 through 9.

Several chapters discuss the needs of different types of telecommuters. Chapter 3 explores the communications requirements of administrative telecommuters to provide the reader with specific examples of *basic voice and data communications applications*, some of the key networking solutions available to address these applications, and the strengths and limitations of each solution. Chapter 4 explores the communications needs of the technical telecommuter to provide the reader with specific examples of *high bandwidth data applications.* Chapter 5 explores the communications needs of the marketing telecommuter in order to provide the reader with specific examples of *mobile, medium data bandwidth and video communications applications.* Chapter 6 explores *computer telephone integration* applications, which are predominantly driven by home agents. This chapter introduces home agents and identifies their communications needs as well as those of call center managers and IS managers. Chapter 7 extends the identification and analysis of communications solutions by focusing on higher level application solutions. Three application solutions are explored in this chapter: groupware, the Internet, and VAN solutions. Chapter 8 explores the key communications *applications of emerging regional telecommuting centers* and associated networking solutions. Chapter 9 explores *emerging telecommunications solutions* that are likely to compete with existing solutions in the next few years.

While Chapters 2 through 9 focus on the macro and micro analyses of the telecommuting trend, Chapter 10 focuses on the supply side. In this chapter, the key network service providers are identified and their current telecommuting services are explored. In addition, potential communications networking business opportunities for these communication service providers are explored.

Chapter 11 reaches conclusions for the demand and supply sides of the telecommuting industry. It summarizes the communications requirements of the various groups of telecommuters and to what degree current and emerging networking solutions can meet these requirements. The chapter then proposes specific courses of action that corporate telecommuting program planners and implementers may want to consider to maximize the chances of success of their telecommuting programs. The chapter is concluded by proposing specific courses of action that communications service providers and application solutions providers may want to consider to maximize the chances of success of their telecommuting programs.

References

[1] Mason, C.F., "Telecommuting Captures the Imagination," *Telephony*, Nov. 23, 1992.

[2] Minoli, D., and A. Tumalillo. *Telecommuting*, Probe Research Corporation Report, June 1994.

[3] "Evolution of the Mobile Professional," *Yankee Watch: Wireless Mobile Communications*, Yankee Group, June 1993.

[4] "Survey: Corporate Mobile Work Force Grows," *Home Furnishing Newspaper*, March 30, 1992, p. 95.

[5] Illingworth, M.M., "Virtual Managers," *Informationweek*, June 13, 1994, pp. 42 ff.

[6] Greig, G., "Office Workers Face a Harsh New World," *The Sunday Times*, England, April 3, 1994, p. 22.

[7] Johnson, K., "In New Jersey, IBM Cuts Space, Frills and Private Desks," *New York Times*, March 7, 1994, p. B1 ff.

[8] "The Virtual Workplace," *Forbes*, Nov. 23, 1992, p. 184.

[9] "Telecommuting and Enhanced Telecommunications," *Yankee Watch: Wireless Mobile Communications*, The Yankee Group, 1993, p. 25.

[10] Hildebrand, C., "Little Folks Wield New Clout in PC Market," *Desktop Computing*, CW, p. 45.

[11] "Telecommuting and Enhanced Telecommunications," *Yankee Watch: Wireless Mobile Communications*, The Yankee Group, p. 14.

[12] Krechmer, K., Action Consulting, "Special Report: Modems—Renewed, Revitalized, Ready!" *Data Communications*, June 1991, p. 88.
[13] "Ameritech Tests ISDN in Telecommuting Pilot Program For Customer Service Reps," *Computerworld*, May 11, 1992, p. 74.
[14] "Telecommuting Gains as Cultural Barriers Fall," *Crain's New York*. Oct. 12, 1992, pp. 26 ff.
[15] "Work-At-Home Services Most Popular, Says Survey," *Telephone Engineer & Management*, Jan. 15, 1993, p. 21.
[16] Minoli, D., "Telecommuting Issues and Solutions," *Managing Data Networks*, Datapro Report 1070, Jan. 1994.
[17] Minoli, D., and M. Vitella, *Cell Relay Service and ATM in Corporate Environments*, New York, NY: McGraw-Hill, 1994.

Telecommuting Demand Characteristics

2

This chapter provides a macroanalysis of the telecommuting demand, including demand side, growth, and distribution. Demand distribution is examined along several dimensions, including geography, company size, and vertical industry. The chapter then identifies telecommuting demand drivers and inhibitors and provides specific examples of telecommuting programs and trials that have been implemented in the last few years. The chapter is concluded by identifying the key corporate players involved in telecommuting, including (1) the corporate decision makers who are involved in the planning, development and implementation of corporate telecommuting programs; and (2) the key groups of telecommuters.

2.1 TELECOMMUTING DEMAND SIZE AND DISTRIBUTION

Estimates of the telecommuting demand size vary among telecommunications industry forecasters. These forecasts, however, have one thing a common: they reflect a large and growing demand for telecommuting. According to industry forecasts, between four and nine million corporate American employees worked at least eight hours per week from home as company employees during business hours in 1992 [1,2]. Overall, more than thirty million people work on their own from home or as company employees during business hours, and an additional nine million company employees work from home after business hours. The emphasis of this book is on company-sponsored telecommuting programs affecting the four-to-eight million people who work from home as company employees. Telecommuting market growth rates are expected to range from 20% to 30% per year over the next five years [2].

Telecommuting demand differs from one region to another. As Figure 2.1(a) shows, the South [2] has the largest number of telecommuters. The Northeast and the West follow, each accounting for 23% of all telecommuters.

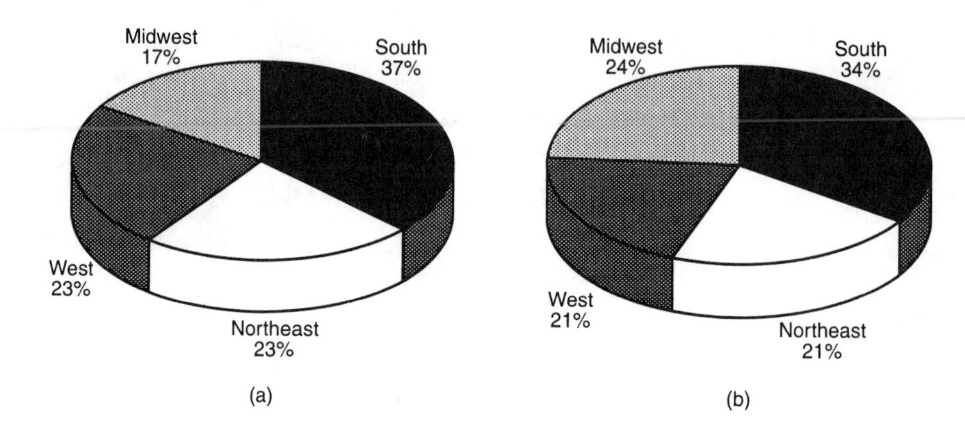

Figure 2.1 (a) Percentage of telecommuters by U.S. region, and (b) population distribution by U.S. region. (*After:* [2].)

The Midwest has the lowest percentage of telecommuters, representing 17% of all telecommuters.

Telecommuting demand also differs from one type of vertical industry to another. Most of the telecommuters are in the personal and business service industry [2], representing over 45% of all the telecommuters. Manufacturing and distribution are in second place, representing over 38% of all the telecommuters. The telecommunications industry is in third place, representing over 12% of all telecommuters. The financial services industry is in fourth place with 9% of all telecommuters. Table 2.1 provides examples of vertical industry-specific corporate telecommuting programs that have been implemented by companies in a variety of industries, including manufacturing, telecommunications, retail, financial services, federal, state, and local governments. As Table 2.1 demonstrates, the scale of these telecommuting programs varies widely, with the number of participants in these programs ranging from 36 to over 10,000 telecommuters.

Telecommuting demand also varies by company size. Large companies employ less than 20% of the telecommuting population. The remaining telecommuters are employed by companies with less than 100 employees.

2.2 TELECOMMUTING DEMAND DRIVERS AND INHIBITORS

The large demand for telecommuting services and its anticipated rapid growth are fueled by nine main drivers: (1) the emergence and growth of the information-based economy, (2) a growing outsourcing trend, (3) government regulations, (4) the drive towards the information superhighway, (5) employee-related factors, (6) employer-related factors, (7) rapid advances in information systems technologies,

Table 2.1
Examples of Telecommuting Programs

Company	*Starting Date*	*Industry*	*Number of Participants*	*Functions*	*Equipment Provider*	*Special Services*
Ameritech	1992	Telecommunications	< 100	Customer Service	Employer	ISDN
Bell Atlantic	1992	Telecommunications	10,000	Managers	Employer	IQ services voice mail
Pacific Bell	1984	Telecommunications	2,000	Agents	Employees mostly	calling card ACD*
JC Penney	1992	Retail	> 200	Telemarketers	Employer	ACD*, terminals, phone calls
Nationwide Insurance	1990 Pilot	Financial Services	36	Agents	Employer pays for full-time employees	Phone line & equipment
Travelers Insurance	1986	Financial Services	200	Agents	See above	ACD*
Apple Computer	1992	Manufacturing	> 1,200	Technical information services, R&D	Employer mostly	Remote access software
State of Washington	1990–92	State government	250+	State employees	Employees	—
Los Angeles County	1989	Local government	> 1,200	Local gov't employees	Employees mostly	—

ACD = Automatic Call Distributor

(8) government benefits, and (9) the recent natural disasters that have affected many parts of the United States. Each of these factors is examined next.

2.2.1 The Emergence of the Information-Based Economy

In the United States as well as in other advanced industrialized countries there is a strong trend toward an information-based economy. This economy is increasingly replacing the manufacturing-based economy that drove the job creation in the United States throughout the 19th and most of the 20th centuries, and will be the key driver of job creation in the next century. At some point in the future only a few percentage points of the total workforce will be in the manufacturing segment (at least in the United States). In 1991, R.B. Reich proposed three categories of jobs in an information-based society: routine production services, in-person services, and symbolic-analytic services. Routine production services entail simple repetitive tasks performed for material transformation or information processing. These jobs account for about 25% of the total U.S. jobs. These jobs, which include data processing, may be candidates for telecommuting. The second category, in-person services, accounts for 30% of the total U.S. jobs; these jobs entail a person-to-person interaction and are not good candidates for telecommuting. The third category includes all problem identifying, problem solving, strategic decision-making, and brokering activities; information system professionals, software developers, research scientists, bankers, lawyers, and so forth, belong to this category. These jobs account for about 20% of the U.S. jobs. Since symbolic analysts rarely need to come in direct contact with the recipient of the work (or this can be done by an initial meeting and follow-up electronic communication), many of these jobs are candidates for telecommuting. Telecommuting can free the workforce from the constraints of time and space [3].

2.2.2 The Growing Outsourcing Trend

In addition to the movement to an information-based economy, there is a related trend toward outsourcing [4,5]. Outsourcing, the turning over of information systems and/or communication functions as a whole or in part to a third-party contractor, is a newly reintroduced solution to the challenge, problem, and expense of creating and running a corporate information enterprise. Outsourcing is a new term for an old concept and is not limited to data processing: it means contracting out work. The consideration is basically simple: if an outside party can do the work more efficiently and inexpensively than can the organization, then the outside party ought do it; if the organization's employees can do the job in a more cost-effective way, then the work ought to remain in-house. As a result of the general contraction of businesses due to a changing economy, corporations are "farming out" functions to various service firms and individuals, ranging from legal to accounting to food service to security to cleaning and to information technology. In addition to more general forms of telecommuting, outsourcing plays a potential role in telecommuting as some aspects of corporate computing (e.g., planning, forecasting, remote management, held desk, customer reps [5]) are severed from the corporation and transferred to the homes of the symbolic analysts.

2.2.3 Government Regulations

In 1990, the Federal Clear Air amendments were passed. This was followed by state government amendments in the early 1990s. For example, in 1992 New Jersey [6] passed the New Jersey Congestion and Air Pollution Act. The

Federally mandated law requires employers with 100 or more employees who work in 18 of New Jersey's 21 counties (Atlantic City, Cape May, and Warren Counties are exempt) to implement an employer trip reduction program (ETRP). This program requires employers to reduce the number of commuter vehicle trips to their locations between 6 a.m. and 10 a.m. To ensure the enforcement of this program, the NJ Department of Transportation (DOT) imposed specific timelines during which employers must comply with the law. Employers have to submit plans to the state by November 1994, providing information about their intentions to meeting trip reduction goals by November 1996, documenting the initial average passenger occupancy (APO) at each work location as of 1994, and describing actions and reporting them to DOT by November 1994. Similar clean air amendments are being adopted in other states, particularly states with populated and polluted urban centers, such as New York City, Houston, and Los Angeles.

In response to the clear air amendments, employers are taking several initiatives including encouraging their employees to rideshare in car pools and van pools, to increase their reliance on public transportation where available, and to work compressed workweek schedules. Employers are also increasingly relying on various types of technologies (e.g., videoconferencing and teleconferencing), and implementing telecommuting programs.

To further encourage corporations to telecommute, several additional government regulations have been introduced over the past few years, notably the National Emergency Legislation Strategy, the Internodal Surface Transportation Efficiency, and the American Disabilities Act.

2.2.4 The Drive Towards the Information Superhighways

A recent U.S. Department of Transportation study showed that there are benefits in a national information network to support, among other functions, telecommuting [7]. As powerful information superhighways are deployed, it becomes easier to move information in an effective manner. Additionally, telecommuting may provide some of the traffic needed to justify the deployment of the emerging national information infrastructure.

2.2.5 Employee-Related Factors

Several surveys indicate that an increasing number of employees are interested in telecommuting. For example, according to a survey of federal computer users conducted by *Government Computer News* [8], 83% of the 500 respondents would work via computer at home or in a near-home remote office if they could. Only 14% say that the opportunity is available to them in their agency. Another survey conducted by Theodore Barry & Associates and Economic Technology Inc. for the New York Department of Public service indicated that the respondents, who were asked to rate eight potential new telecommunications services, expressed the most interest in the work-at-home group of services.

Employees are increasingly requesting from their employers the right to telecommute to spend more time with their spouses and children rather than in traffic; to save the money that they spend on purchasing, maintaining, and fueling a car, public transportation, and summer and winter clothes. Employees also believe that as telecommuters they can be more productive because they can work with no interruptions from their coworkers and in more comfortable surroundings. In addition, an increasing number of employees would like to work at home to reduce their exposure to air pollution, particularly in major urban areas, such as Los Angeles, and to reduce stress associated with commuting. Those employees are expressing these views to their supervisors.

A recent market research study prioritized the benefits of working at home from a telecommuter's point of view. As Table 2.2 shows, the highest priority of telecommuters and other corporate workers was the desire to get more work done.

Table 2.2
Reasons Cited by Telecommuters for Working at Home

Reasons	*Percentage of Telecommuters*
Get more work done	33%
Earn more money	25%
More flexible schedule	21%
Be my own "boss"	16%
Start own business	14%
Family care	14%
Reduce commuting	8%

2.2.6 Employer-Driven Factors

Faced with increasing U.S. and international competition, regulatory pressures from the federal, state, and local governments, and increasing demands from their employees, employers are increasingly developing and implementing telecommuting programs to increase employees' productivity. These productivity gains can be accomplished by telecommuters because they can work longer hours (6 to 10 hours per week) in compensation for eliminated commuting, they can improve work quality due to easier management of distractions, and they have stronger work incentives resulting from a greater sense of being in control.

Employers are also developing and implementing telecommuting programs to reduce recruiting, relocation, and health insurance costs; to lower overhead costs associated with office space by making double use of desks; to reduce retraining costs resulting from employee turnover of up to 33%; and to recover from disasters more quickly by establishing communications links between corporate offices, including corporate data center locations and the telecommuters' homes so that in case of a major disaster, corporate operations could be resumed rapidly. Other sources of cost savings to employers include reduction in travel costs and reduction in penalties associated with noncompliance with government air quality mandates, which translates into less cost associated with penalty fees. Telecommuting also enables employers to reduce cost by utilizing untapped resources, including over 30 million disabled Americans who could join the workforce through solutions such as telecommuting.

Table 2.3 quantifies the benefits of telecommuting to employers. These numbers are based on estimates of Fleming LTD [9], telecommuting advisors to the State of California and other employers. Fleming LTD utilizes these numbers to demonstrate the benefits of telecommuting to employers. These numbers are based on the following assumptions: (1) annual salary of $20,000, (2) productivity increase of 20%, (3) reduced personnel cost of 10%, (4) parking at

Table 2.3
Benefits of Telecommuting to Employers

Personnel Factors	*Benefit-Savings*
Productivity (quantity)	$3,000
Productivity (quality)	$1,000
Recruiting and training	$1,000
Retention/absenteeism	$1,000
Facilities	
Office space	$1,800
Parking	$200
Annual total per employee	$8,000

$400 per year reduced by 40%, and (5) use of central facilities of 150 square foot rent per year reduced by 40%.

2.2.7 Advances in Information Systems Technologies

The continued introduction of computing technology in the corporation, in particular distributed computing systems, as well as the increased availability of high-quality/high-capacity switched digital communication services, are making telecommuting possible, as well as cost-effective, for all concerned. This "electronic office" enables the remote performance of many business functions.

2.2.8 Benefits to State and Local Governments

Government support for telecommuting programs has also been significant. This support has been driven not only by their desire to reduce air pollution, but also by their desire to (1) reduce traffic congestion, resulting in fewer accidents; conserve energy; (2) reduce neighborhood crime by increasing daytime population; (3) create employment for disabled workers; and (4) reduce expenses. States that support a telecommuting program can also benefit from lower cost of maintaining highways. For example, according to a report from the California Engineering Foundation entitled "Transportation Redefined," telecommuting could help California save about $5 billion annually in fuel costs and productivity. Telecommuting could accomplish these savings by easing the strain on California highways.

2.2.9 Natural Disasters

The United States has witnessed a number of natural disasters. For example, both Los Angeles and San Francisco recently suffered a number of consecutive earthquakes. Natural disasters are likely to drive the employers to adopt telecommuting programs so that they could avoid lengthy work disruptions in case of a natural disaster.

Retarding Factors

While the factors mentioned above are driving the growing telecommuting trend, two factors are retarding its growth:

1. *Management cultural barriers.* Many supervisors have a traditional mindset: they use clock management as a tool to measure employee performance. As competitive pressures increase, supervisors will have to change their mindset from a "clock management" orientation to a "management by results" orientation. [6,11,12].

2. *Union resistance to the work-at-home concept.* Several major unions are opposed to the telecommuting concept because they are concerned about the possibility of the establishment of electronic sweatshops [13].

The factors driving the growth of the telecommuting market are expected to overcome the obstacles that are retarding its growth. Given these market dynamics, the telecommuting market is expected to grow at 20% to 30% per year for the next few years.

2.3 CORPORATE TELECOMMUTING PROGRAM PARTICIPANTS

As indicated in Section 2.1, corporate telecommuting programs are implemented predominantly by smaller firms (representing 80%) as well as by larger corporations. These implementations are either through informal or formal arrangements. Decisions associated with informal arrangements usually involve the telecommuter and his or her supervisor. These arrangements are made in smaller or larger corporations. In contrast, formal telecommuting programs are usually adopted by larger corporations.

Several corporate functions are involved in the process of developing and implementing telecommuting programs. Some of those performing these functions are also likely candidates for participation in corporate telecommuting programs. To identify the corporate functions that are likely to participate in a telecommuting program, the value chain model developed by Michael Porter, Harvard University, has been utilized. This model is generic in nature and applies to any productive unit whether this unit produces and sells products or services. This model divides corporate activities into two major groups:

1. *Primary activities,* including inbound logistics, outbound logistics, marketing and sales, operation and after-sales customer service.
2. *Support activities,* including upper management (legal, finance, and strategic planning), human resources management (HRM), procurement, and technology development. Technology development encompasses corporate employees that develop and implement technical solutions in support of the primary activities described below. Examples of these technical developers include those performing R&D functions and MIS functions. These activities cut across multiple primary activities as shown in Figure 2.2.

Three support activities—human resource management (HRM), upper management, and MIS—are involved in planning and implementing a corporate telecommuting program. The role of these activities is described next.

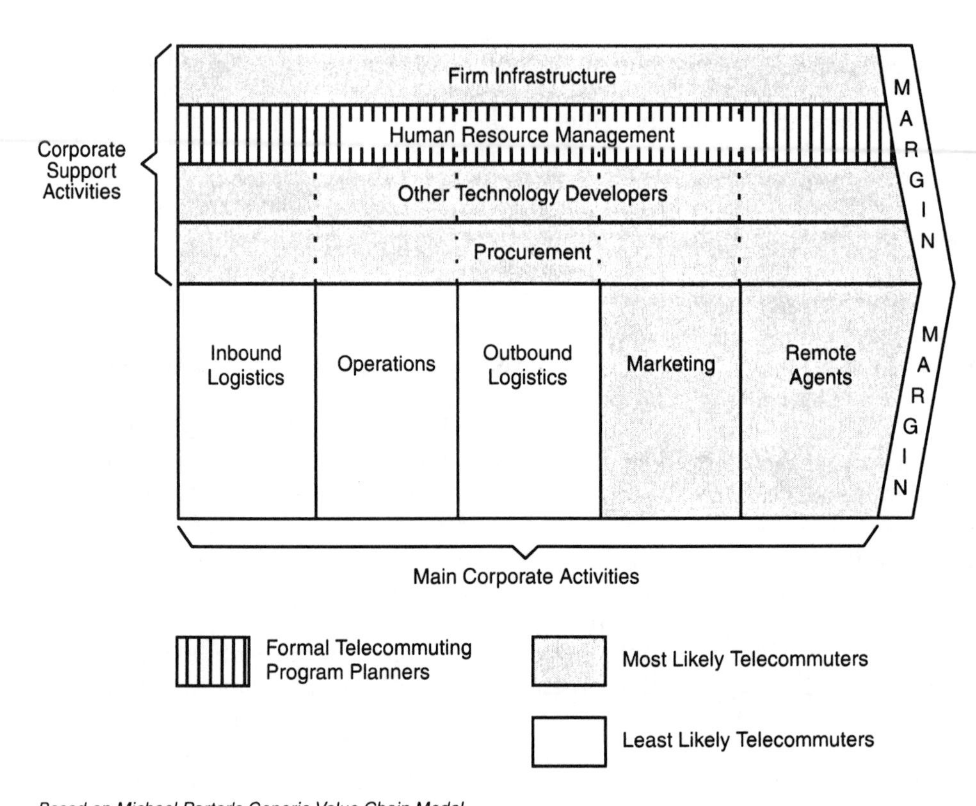

Figure 2.2 Corporate telecommuting participants: the generic value chain.

2.3.1 The Role of Human Resource Management

A human resources management department, particularly in medium to large businesses, plays a central role in the planning of formal corporate telecommuting programs. These are programs that are adopted by large corporations. This department defines the scope of these programs. In most corporations, formal telecommuting programs start on a small scale as pilot programs. They may then be further expanded following the evaluation of the initial results. The human resources department may also (1) plan telecommuting programs, (2) determine the number of participants in these programs, (3) select the functions and the corresponding departments that can participate in these programs, and (4) select the individual participants in this program. This selection is made in consultation with the supervisors of these departments. In addition, this department quantifies the benefits of telecommuting and presents the results to upper management for final approval. While human resource managers play a key role in the planning of formal corporate telecommuting programs, they may play a minor role in smaller organizations (with fewer than 100 employees).

2.3.2 The Role of Information Systems

Two corporate information systems (IS) departments may become involved in the implementation of a corporate telecommuting program: the data processing department and/or the communications (voice and/or data) department. The data processing department becomes involved if the telecommuting program involves a large purchase of computers or if secured access to corporate data processing computers is required by telecommuters. The communications department becomes involved in the selection and management of the technologies that are required to establish communications links between corporate offices and the telecommuters' homes, particularly if establishing these links requires the subscription to a new service, such as ISDN.

2.3.3 The Telecommuters

As Figure 2.2 shows, the least likely candidates for participation in a telecommuting program are those who perform operations functions (e.g., manufacturing quality control) or inbound and outbound logistics functions (e.g., inventory control). The nature of these functions demands the availability of those performing these functions at corporate sites during business hours.

Figure 2.2 also shows that the potential telecommuters are those corporate employees who perform functions that do not require their presence in the corporate office either full-time or part-time. These employees may be (1) administrative telecommuters, (2) technology developers, (3) sales and marketing, or (4) remote agents. Each of these groups of telecommuters performs a number of functions, which are described below.

Administrative telecommuting functions encompass several support activities including human resource management, upper management, and legal, financial, and procurement functions, as seen in Figure 2.2. (A more detailed description of administrative telecommuters and their inter and intracompany communications requirements is provided in Chapter 3).

Technical telecommuters include those who develop and implement technical solutions in support of internal corporate functions (e.g., inbound logistics, operations) and in support of their clients. (A more detailed description of the technical telecommuters and their intracompany communications requirements is provided in Chapter 4).

Marketing telecommuters include those who perform market research, planning, promotion, and development. (A more detailed description of marketing telecommuters and their inter and intracompany communications requirements is provided in Chapter 5).

Remote agents include outbound and inbound telemarketers, market surveyors, help line operators, customer service representatives, and order entry clerks. (A more detailed description of remote agents and their communications requirements and solution is provided in Chapter 6).

References

[1] Mason, C.F., "Telecommuting Captures the Imagination," *Telephony*, Nov. 23, 1992.
[2] Minoli, D., and A. Tumalillo, *Telecommuting*, Report by Probe Research Corporation, June 1994.
[3] Shae, Patricia A., "Telecommuting—Freeing the Workforce From Constraints of Space and Time," *Answer*, May 1992, p. 50 ff.
[4] Apte, U., "Global Outsourcing of Information Systems and Processing Services," *The Information Society*, Vol. 7, 1990, p. 287 ff.
[5] Minoli, D., *Reengineering and Outsourcing IS and Communication Functions: Methods and Benefit Analysis*, New York, NY: McGraw-Hill, 1994.
[6] Welsh, D., "Telecommuting Will Require Change in Management," *Network World*, April 12, 1993, p. 40.
[7] "DOT Study Shows Benefits of National Information Network," PHONE+, June 1, 1993, p. 17.
[8] "Feds Want to Keep Their Work Closer to Home," *Government Computer News*, Aug. 17, 1992, p. 1.
[9] Kocher, C. E. "Bud," "Pinpointing the Work-At-Home Market," *Telephony*, Feb. 8, 1993, p. 26.
[10] "Telecommuting Gets Vote of Confidence," *Network World*, Aug. 31, 1992, p. 29.
[11] "Telecommuting Obstacle: Management, Not Technology," *Telemanagement*, Oct. 1992, p. 8.
[12] "Telecommuting Gains as Cultural Barriers Fall," *Crain's New York*, Oct. 12, 1992, p. 26 ff.
[13] "Union Resistance Could Slow the Growth of Telecommuting," *Wall Street Journal*, Sept. 22, 1992.

Basic Communications Applications for Administrative Telecommuters

3

Most administrative telecommuters have basic communications requirements. This chapter explores these requirements to provide the reader with specific examples of basic voice and data communications applications, some of the key networking solutions available to address these applications, and the strengths and limitations of each solution.

3.1 WHO ARE THE ADMINISTRATIVE TELECOMMUTERS?

Administrative telecommuters are the largest group of telecommuters, representing more than 47% of the corporate telecommuting population. This group performs all corporate support activities with the exception of technology development. The group includes corporate executives, lawyers, accountants, financial managers and analysts, and procurement managers.

3.2 NETWORKING REQUIREMENTS OF ADMINISTRATIVE TELECOMMUTERS

Administrative telecommuters have several voice and data communications requirements. Voice requirements include: (1) access to voice mail to retrieve messages from the office; (2) access to corporate Centrex/PBX system to utilize voice communications features, such as call handling; (3) the ability to confer with supervisors, suppliers, customers, and other employees on a short notice through conference calling; and (4) the ability to screen incoming calls to cut down on interruptions. In addition, administrative telecommuters, such as corporate executives, would like to have the messages that they receive from their offices transparently rerouted to their homes to ensure that they do not miss any important calls.

Data communications requirements of administrative telecommuters include access to departmental LAN servers to (1) retrieve local files, such as personnel, financial, supplier, and legal files; (2) access departmental databases; (3) access e-mail; and (4) access shared facilities (such as printers and modem pools). Administrative telecommuters also need to access host computers to access (a) corporate databases, which may include information about suppliers and legal rulings; (b) corporate e-mail systems to exchange messages with other corporate employees or with external suppliers; and (c) access shared facilities (printers and modem pools).

3.3 NETWORKING REQUIREMENTS OF THE INFORMATION SYSTEMS MANAGERS

IS managers have their own set of criteria in selecting a networking solution and in supporting the networking requirements of telecommuters. These criteria not only apply to administrative telecommuters, but to all the other groups of telecommuters as well. IS managers take into account (or should take into account) the following criteria in selecting network solutions.

1. *Quality of service*: The quality of provisioning is an important factor that reflects the overall quality of a service. The quality of provisioning is a function of the time required by the communications service provider to provision a service and the degree of simplicity associated with the delivery of the service. Another factor that reflects the quality of a service is the performance of the service, which is a function of several factors such as the ability of a service to guarantee the delivery of the nominal bandwidth specified by the service provider and the range of bandwidth supported by the service.
2. *Availability of the service*: A major factor that reflects the availability of the service is the number of points of presence (POPs) that support this service. The number of POPs supported by each provider has a direct bearing on the cost of dedicated local-loop access. Another factor that reflects the long-term availability of a service is the degree of vendor support. A third factor that reflects the availability of the service is the degree of support for the service by the various types of vendors, including equipment vendors, local exchange carriers (LECs), and interexchange carriers (IXCs). In addition, IS managers are increasingly seeking the option of an alternate access provider (AAP). A fourth factor reflecting the availability of the service is the degree of its international presence. This factor is becoming increasingly important as a greater number of companies, irrespective of their size, are competing in international markets. In an international context, *telecommuting* means "stay

in your own country, but work for a U.S. firm." For example, software developers stationed in Asia write code that resides on U.S. servers.

3. *Transport and interface standards*: The standards may be either formal or defacto.
4. *Low cost*: Cost minimization is another important factor in selecting networking solutions that minimize the cost of equipment and both nonrecurring and recurring charges. One of the key factors that drive this cost is the number of access lines required to support the range of telecommuting applications. To minimize this cost, IS managers are increasingly seeking *integrated access* solutions that enable them to consolidate the traffic originating from various telecommuters onto a single access line at the corporate office. Another service-cost driver is the changes that an IS manager needs to make in his or her existing networking infrastructure to support a new service. This infrastructure includes the installed networks, wiring, and corporate-based equipment. IS managers take into account their existing infrastructure in the development of the IS plans. For example, if an IS manager selected and implemented a frame relay network, he or she would prefer to select a telecommuting networking solution that is compatible with the frame relay network.
5. *Secure communications*: Some IS managers may need security solutions that are network-based. For example, encryption may be required.
6. *Reliability*: Several factors reflect the reliability of a service, including its ability to retain POTS during power outage, its ability to support the operation of voice equipment during power outage, and the degree of development of existing telephone operations supporting the service.

3.4 NETWORKING SOLUTIONS

In establishing networking links between their homes and corporate offices, administrative telecommuters may rely on one of the following options: (1) dialup/modem solutions, (2) ISDN solutions, or (3) data over voice (DOV) solutions.

3.4.1 The Dialup/Modem Solution

This solution is the present mode of operation (PMO). As Figure 3.1 shows, to implement this solution administrative telecommuters rely on (1) an analog phone, (2) a data terminal connected to a 9.6 Kbps or higher speed modem, and (3) one POTS lines supporting voice and data applications.

At the corporate side, a modem pool is required to support these telecommuters. These modems are interconnected to a multiport terminal server, providing the administrative telecommuters with access to LAN resources.

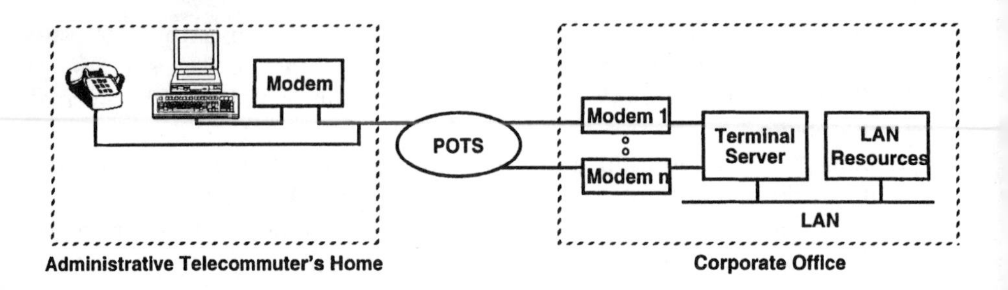

Figure 3.1 POTS-based administrative telecommuter/corporate office links.

This solution is popular because it supports POTS-based features; it is ubiquitous; it does not require any major changes to existing home or corporate-based equipment; it supports voice equipment operation during local power outage; it addresses the bandwidth requirements of the administrative telecommuters; and it is supported by established telephone transport and interface standards. In addition, the cost of purchasing equipment is low.

This solution, however, is not ideal for several reasons: (1) standard fallback is not available; (2) if a POTS line fails, it cannot be remotely isolated; (3) modems can step down to lower speeds (e.g., from 28.8 Kbps to 19.2 Kbps or 9.6 Kbps) when POTS lines voice grades are used; and (4) a voice conversation cannot be maintained during a data session.

3.4.2 The ISDN Solution

One of the networking services that is becoming increasingly available in most U.S. regions and internationally is integrated service digital network (ISDN).[1] ISDN is an end-to-end digital communications technology that extends from the telecommuter's premises, through the telephone network, to the user at the far end (e.g., a corporate location). All interfaces to the ISDN network are based on U.S. and international standards. As noted in Chapter 1, ISDN is designed to carry multiple 64-Kbps B channels (1B, 2B, 23B) and one 16-Kbps D channel. In addition, several LECs are trialing ISDN with *N*B capabilities, where *N* is greater than 2 and less than 23. The B channels can be used to deliver a variety of data, voice, and video services, with one service typically dedicated to each B channel. The D channel is a multiplexed message-based channel that is intended to deliver data (possibly several low-speed channels) and signaling (to control the use of the B and D channels).

To facilitate interworking, in 1991, the BOCs, the Corporations for Open Systems International, the North American ISDN Users' Forum (NIUF), and switch vendors agreed to adopt Bellcore-defined ISDN generic requirements,

1. For a detailed treatment of ISDN, refer to Minoli, D., *Enterprise Networking: Fractional T1 to SONET, Frame Relay to BISDN*, Norwood, MA: Artech House, 1993.

which include three phases. The first phase of the standards is national ISDN-1 (NI-1), which provides a basic ISDN platform supporting multiple basic rate interface (BRI) services. National ISDN-2 (NI-2) standardizes primary rate interface (PRI) and supports terminal portability. NI-3 is in the planning stage. The specifications of NI-3 will bring greater ISDN interoperability with frame relay and personal communications services [1].

Deployment of ISDN by local and exchange carriers has been accelerating. As Figure 3.2 shows, the Bell operating companies are aggressively deploying ISDN. By 1995, the percentage of lines with access to ISDN will reach 90% in some regions, such as the Bell Atlantic region [2]. In addition to the BOCs, the major IXCs also support ISDN. These IXCs include AT&T, MCI, and Sprint. Equipment vendor support is also on the rise. There are currently more than 200 ISDN products offered by over 60 vendors. Prices of ISDN equipment, initially high, are now coming down.

ISDN tariffs appear to be economical with NYNEX; for example, NYNEX is charging about $40 for a single-line ISDN.

3.4.2.1 ISDN Applications

ISDN is a versatile technology that can address a wide range of applications. Many of these ISDN applications were demonstrated during the transcontinental ISDN project in 1992 (TRIP '92). Many ISDN solutions were also catalogued by the national ISDN user forum (NIUF), and are continually being upgraded [3]. This section examines an ISDN solution that can meet the basic communications requirements of the telecommuter.

ISDN (1B + D)[2] addresses the requirements of administrative telecommuters for *voice communications and light data transmission.* The administrative telecommuter can use the B channel for voice, while packet data is transmitted on the D channel at 9.6 Kbps. Through ISDN (1B + D), an administrative telecommuter can (1) access to corporate e-mail; (2) access corporate directories; and (3) establish secure communications through the ISDN X.25 fast packet feature supported on the D channel, enabling only a particular set of users to access the system. In addition, ISDN provides the administrative telecommuter with *optional advanced voice features* (supported on the D channel), such as call number identification, which can cut down on interruption by letting the user screen incoming calls [4].

As Figure 3.3 shows, to establish an ISDN link with the corporate office, administrative telecommuters need *a digital telephone* with (1) B-channel capabilities, including voice calling on a single B channel and if using NI-1, support for one directory number and service profile ID; (2) D-channel capabilities,

2. ISDN (1B + D) may not be ubiquitously available, as some RBOCs (e.g., Pacific Bell) are opting to offer ISDN (2B + D) instead at reasonable prices (e.g., $28). For information about ISDN (2B + D) solutions, please refer to Chapters 4, 5, and 6.

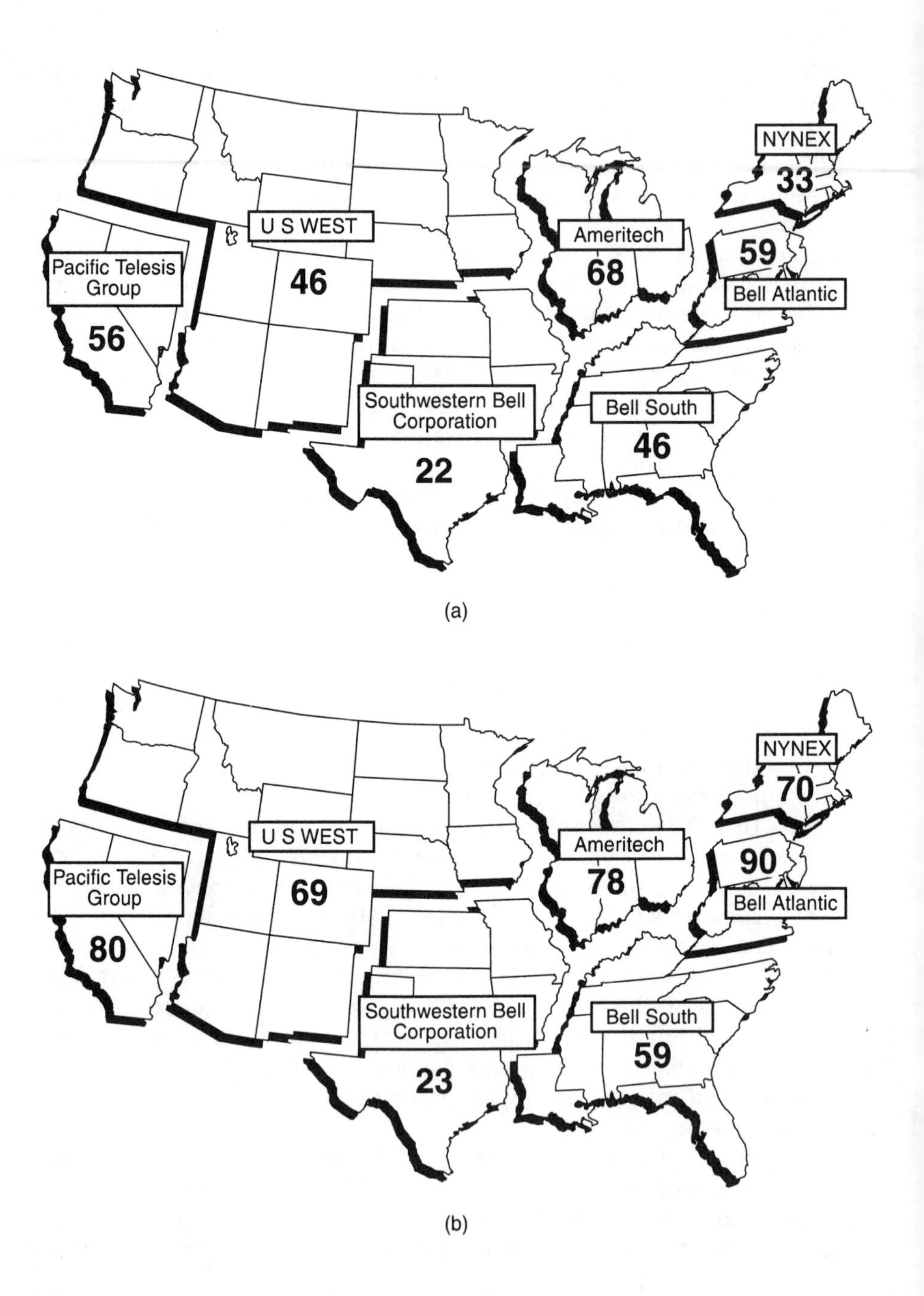

Figure 3.2 (a) Percentage of lines with access to ISDN in 1993, and (b) projected percentage of lines with access to ISDN in 1995. (*After*: [2].)

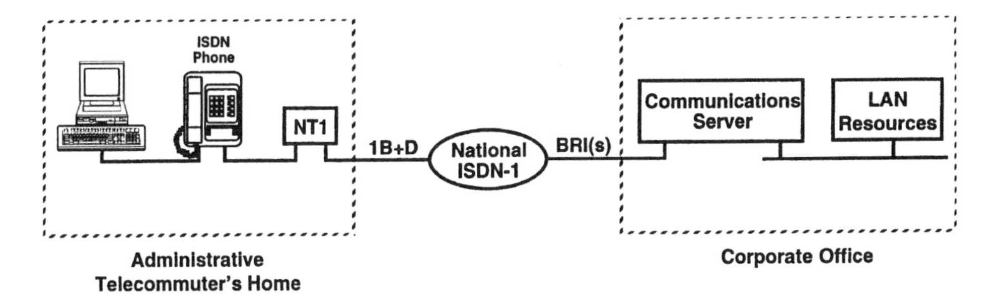

Figure 3.3 ISDN-based administrative telecommute/corporate office links.

including support for PAD functions (X.3, X.28, and X.29), and support for various data features, including at least one two-way logical channel, optional fast select and fast select acceptance, optional reverse charging, optional throughput class negotiation, and optional flow control parameter negotiations.

Administrative telecommuters also need a personal computer equipped with applications software, optional communications software and interfaces to an ISDN phone (e.g., RS 232). Both the ISDN telephone and the PC are linked to the ISDN network through an NT-1.

On the corporate side, a communications server with multiple BRI capabilities is required to support multiple telecommuters. This communications server must also have: (1) applications software that is compatible with communications software and with telecommuters' software; and (2) B-channel capabilities for each BRI (including packet data on each B channel), support for multiple packet data features (including multiline hunt group, with hunt DN, unrestricted fast packet, and fast select acceptance), as well as reverse charge acceptance.

ISDN offers the telecommuter several benefits over the dialup/modem solution: (1) it guarantees the delivery of information at the bandwidth promised to the telecommuter; (2) it provides multiple call management services (call waiting, call forwarding, calling number ID), which are transmitted on the D channel while the B channel handles voice; (3) the administrative telecommuter can receive and make phone calls during data communications sessions; (4) national ISDN-2 standards are providing standardized PRI solutions that will enable an IS manager to concentrate data traffic originating from the administrative telecommuters' homes at the corporate site; and (5) this solution provides a migration path to higher bandwidth, such as 128 Kbps with 2B.

ISDN, however, has several limitations. First, ISDN does not provide loop-backup powering during a local power outage. Second, some rewiring of telecommuter premises will be needed to install ISDN. Finally, ISDN has higher equipment costs than POTS. The ISDN telephone set costs $200–$400, while the cost of an analog telephone cost $50–$80. In addition to the cost of the ISDN phone, the corporation will incur the cost of NT1 ($250), and ISDN ports ($350). This could be a barrier for the providers of ISDN services if the administrative telecommuter has to purchase his or her own equipment instead of relying on the corporation.

3.4.3 The Data-Over-Voice Solution

The term data over voice (DOV) refers to a group of existing technologies that enable the transmission of data and voice over the same copper pair. Data and voice are sent in separate channels that operate in different frequency ranges. The most common use of this technology is for information services.

DOV currently exists in three forms: (1) analog voice and data; (2) analog voice, digital data; and (3) digital voice and data. Each of these alternatives is examined next.

3.4.3.1 Analog Voice and Data

Analog voice bandwidth ranges from the 300 Hz–3.1 kHz bandwidth while analog data bandwidth is above 10 kHz. The modulation of these frequencies varies from one vendor to another. The multiplexed data and voice service are transmitted over a single access line to a central office (CO)-based multiplexer where they are demultiplexed (Figure 3.4). These CO-based multiplexers have to be identical to those at the administrative telecommuters' homes.

3.4.3.2 Analog Voice, Digital Data

Voice transmission is the same as in the above discussion. The data is transmitted as encoded pulses in a higher frequency band using various modulation techniques. The technology supports synchronous and asynchronous digital data transmission up to 19.2 Kbps.

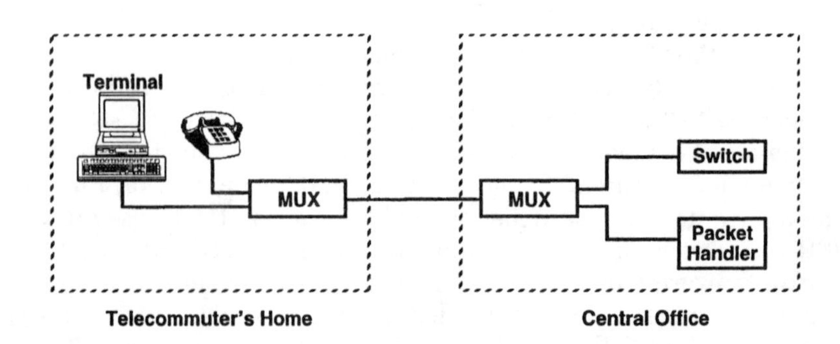

Figure 3.4 The DOV solution.

3.4.3.3 Digitized Voice, Digital Data

The voice channel is digitized and digitally multiplexed with the data channel. The data stream is demultiplexed into the voice and data channels at the receiver end. If the digital transmission fails for any reason, the facility automatically returns to metallic voice transmission.

The DOV solution has several advantages: (1) it supports POTS voice, signaling, and CLASS features; (2) it delivers POTS and supports voice equipment operation during local power outages; (3) it provides the administrative telecommuter with simultaneous access to voice and data services; (4) it does not require rewiring for voice services; (5) it meets the bandwidth requirements of the administrative telecommuters (if needed, more bandwidth is available for this technology to support additional channels); and (6) its home-based equipment cost is low and it can be ordered today.

This solution, however, is not ideal because (1) neither a formal or a defacto transport standard exists; (2) it does not provide line fault isolation; and (3) may only be available in a few regions. For example, Pacific Bell announced plans to offer digital data over voice (DDOV) in 1992.[3]

3.5 CONCLUSIONS

The communications requirements of most administrative telecommuters are basic. They include *voice communications and casual data usage*. Three of the networking solutions that can meet these requirements are: dialup/modem solution with one line, ISDN (1B + D), and DOV. Table 3.1 summarizes the strengths and limitations of each solution.

Table 3.1
Comparison of Administrative Telecommuter Networking Solutions

Customer Selection Criteria	*Networking Solution*		
	POTS (One Line)	*ISDN (1B + D)*	*DOV*
Applications supported:			
Small file transfer	•	•	•
Voice	•	•	•

3. This information is based on published reports; although it is believed to be correct, it has not yet been verified with the carriers.

Table 3.1 (continued)
Comparison of Administrative Telecommuter Networking Solutions

Customer Selection Criteria	*Networking Solution*		
	POTS (One Line)	*ISDN (1B + D)*	*DOV*
Bandwidth		•	
Standards	•	•	
Reliability		•	
Availability	•	•	

References

[1] "ISDN Train Keeps A Rollin'," *Telephony*, May 31, 1993, p. 18.

[2] "ISDN Deployment: The Tortoise Makes Its Move," *Data Communications*, Sept. 1993, p. 17.

[3] NIUF Application Analysis Working Group, *A Catalog of National ISDN Solutions for Selected NIUF Applications*, second edition, Feb. 11, 1994.

[4] "Telecommuting with Casual Data Requirements," *A Catalog of National ISDN Solutions for Selected NIUF Applications*, second edition, Feb. 11, 1994, pp. 3-67 to 3-69.

[5] D. Minoli, *Enterprise Networking, From Fraction T1 to SONET, From Frame Relay to BISDN*, Norwood, MA: Artech House, 1993.

4 High-Bandwidth Communications Applications for Technical Telecommuters

The objective of this chapter is to explore the communications needs of the technical telecommuter in order to provide the reader with specific examples of high-bandwidth data applications.

To accomplish this objective, the chapter introduces the reader to the functions performed by the technical telecommuters and the value they create to their employers. The chapter then explores the communications applications of technical telecommuters, their communications requirements associated with these applications, and the potential networking solutions addressing these requirements. For each networking solution, the technical details are provided and the strengths and limitations are described.

4.1 WHO ARE THE TECHNICAL TELECOMMUTERS?

Technical developers represent the second largest group of telecommuters, accounting for almost 30% of the total number of telecommuters. This group performs technical research, development, and implementation functions in a wide range of industries, such as health care, education, and manufacturing. Examples of these technical telecommuters include engineers (e.g., mechanical, architectural, electronic, and civil); information systems engineers, managers, and technicians; software developers; physicists engaged in studies such as fluid dynamics or turbulence; medical researchers, technicians, and doctors; chemists involved in the analysis of molecular structures, chemical reactions, and statistical mechanics of solids; molecular biologists involved in cladistic analysis; and scientists in fields such as atmospheric sciences, astronomy, and astrophysics. Those telecommuters may perform tests and experiments on company premises and then manipulate, study, or publish the data while working at home.

Technical developers create value to their employers (corporations, hospitals, universities) by designing, developing, and implementing technical solutions. These solutions may be products and/or services that a corporation delivers to its customers. Solutions may also be designed to (re)engineer internal business processes. Internal solutions may either support a single corporate function (e.g., operations, marketing, inbound logistics, or administrative) or multiple functions. Examples of these internal solutions include a document imaging system designed to increase the efficiency of paper workflow processes.

Upper management of corporations are seeking to increase the value created by these technical developers to:

1. Enhance the quality and/or speed of the process of delivering existing and new products to current and emerging customers. These enhancements enable the employers to differentiate themselves from competitive products and services.
2. Improve the cost structure of the corporation through the re-engineering of internal business processes.

To accomplish corporate goals, technical developers, whether they are working at home or in the office, need to establish *rapid, reliable, continuous, and cost-effective* communication links with other corporate employees, such as technical colleagues, sales and marketing employees, and inbound and outbound logistics; and with their technical counterparts working for their trading partners, including corporate suppliers, and customers.

4.2 TECHNICAL TELECOMMUTING APPLICATIONS

Technical developers may spend two to three days per week at home. While working at home, they would like to duplicate the communications environment that they have in the office. Several voice and data applications drive the communications needs of technical telecommuters. Some of the key applications are described below.

4.2.1 Voice Applications

Technical telecommuters (would like to) rely on voice communications to *retrieve voice mail* messages from the office and to *access corporate Centrex/PBX system* to utilize corporate voice communications capabilities, such as call handling features, three-way calling, call transfer, consultation hold, call waiting, and calling number identification. Technical developers would also like to have the messages that they receive from their offices or from their technical colleagues transparently rerouted to their homes.

4.2.2 Data Communications Applications

The data communications needs of technical telecommuters are driven by four applications: (1) electronic mail, (2) large file transfer, (3) remote access to databases, and (4) data conferencing.

1. *Electronic mail* is the most pervasive application of technical developers. Technical telecommuters rely on e-mail to collaborate with other technical developers in other locations, exchanging comments about abstracts, manuscripts, papers, and books. These collaborators reside in corporate locations or in other locations that may be local, across the state, across the country, or overseas. This application is not a major bandwidth driver today. However, corporations are increasingly migrating to graphic client/server applications, enabling end users to attach files that include image-formatted text and audio. This will increase bandwidth requirements of the electronic mail application.
2. *Access to databases*: Technical telecommuters have a need to access a number of databases, such as corporate and departmental databases, which may include a library of engineering objects. These databases may also contain equations that an engineer may rely upon in performing specific design calculations. Technical telecommuters may also need to access public databases. For example, members of the medical and chemical communities need to access databases [1], such as GENBANK, the Cambridge Crystallographic Database, the Human Genome Sequence and the Global Seismic Databank. These databases are accepted standards in their respective communities and are quite large. Consequently, sharing access to these databases among the members of these communities is a more economic solution than duplicating these databases at every employers' location throughout the country and incurring the cost of creating, upgrading, and maintaining these databases.
3. *Large file transfer*: The generation and transmission of images have become key needs in various technical fields [2]. For example, instantaneous transmission of X-ray images from a patient in a hospital to a doctor's office or residence is becoming increasingly required. Another example is offered by engineers who may need to download files, including graphics, CAD systems, or complex software models, from the corporate office. These files are usually shared among multiple engineers who are working on the same projects. Such *imaging applications* are major bandwidth drivers. The typical message length associated with X-rays and CAD/CAM imaging is 10 MB [2]. Scientific files (for example, molecular dynamics) are also major bandwidth drivers. The typical message length associated with this application is 80 KB to 3 MB [3].

4. *Data conferencing applications*: In the future, technical telecommuters would like to be able to establish computer conferencing sessions with supervisors, partners, and technical colleagues on a short notice to jointly manage projects effectively, including scheduling project time frames and to collaborate with other colleagues (e.g., marketing, sales, or customer service) in the process of developing new products. For example, those team members residing or working in different locations may be working on a design problem, requiring the establishment of a voice and data conference session among them in which product illustrations, specifications (text-based), and graphics are viewed on the computer screens of all the participants, where problems are discussed and resolved in real time. At the end of the conference, the latest version of the product design would be available to all the participants and saved in individual PC files, or in a LAN server. Technical telecommuters would also like to establish computer conferencing sessions to consult with each other. For example, a physician working at home may want to establish long-distance conversations with a physician in another location while simultaneously viewing a patient's diagnostic images.

4.3 NETWORKING REQUIREMENTS OF TECHNICAL TELECOMMUTERS

Technical telecommuters are seeking networking solutions that provide them with access to the remote applications identified above and that meet their three major requirements: high bandwidth, reliability, and low cost.

Technical telecommuters have high bandwidth requirements. As mentioned above, imaging applications are among the major bandwidth drivers. Technical telecommuters are seeking networking solutions that provide them with *reliable uninterrupted* access to the applications mentioned above during business hours as well as after business hours. Technical telecommuters may actually need reliable networking support 24 hours a days for two reasons: (1) the creative process associated with technical development does not know any particular hours, and (2) their technical partners may be working in other countries in different time zones. In addition, technical telecommuters are seeking low-cost networking solutions if they are required by their employers to incur the partial or total cost associated with working at home. This cost includes the cost of equipment (e.g., computer, modem, or terminal adapter and one telephone line). The cost distribution between the employer and the employee depends upon the size of the employee's organization, the perceived benefit by the employer of the value of working at home, and whether the telecommuting arrangement is formal or informal.

4.4 NETWORKING SOLUTIONS

IS managers and technical telecommuters may currently have several networking solutions that meet at least some of their requirements. Three groups of solutions are examined in this chapter: (1) POTS with two lines, and (2) ISDN solutions. Each of these groups is also compared according to the network selection criteria of IS managers and technical telecommuters.

4.4.1 The Modem/Dialup Solution

The modem/dialup solution is the predominant present mode of operation (PMO). As Figure 4.1 shows, this solution can support *voice communications, low-bandwidth data file transfer, data conferencing,* and *electronic mail.*

To implement this solution, technical telecommuters need a high-end workstation or PC. Technology developers use their high-end PCs or engineering workstations to develop software or to perform computer-aided design and computer-aided engineering (CAD/CAE). Technical telecommuters also need a high-speed modem operating at a speed of 14.4 Kbps with a compression ratio of 3:1. V.34 modems now support a basic rate of 28.8 Kbps; some models support compression rates of 8:1, achieving a nominal throughput of 230 Kbps [4]. This modem is either directly attached to the workstation or is an integral part of the workstation. In addition, technical telecommuters need two POTS lines, one should be dedicated to voice transmission to provide the technical telecommuter with access to Centrex services, while the other line should be dedicated to data transmission.

At the corporate side, a pool of LAN adapters is dedicated to supporting the individual telecommuters. These adapters provide the technical telecommuter with access to LAN resources, such as e-mail and directory services.

This solution is the current predominant solution that a technical telecommuter relies upon in establishing communications links with the corporate

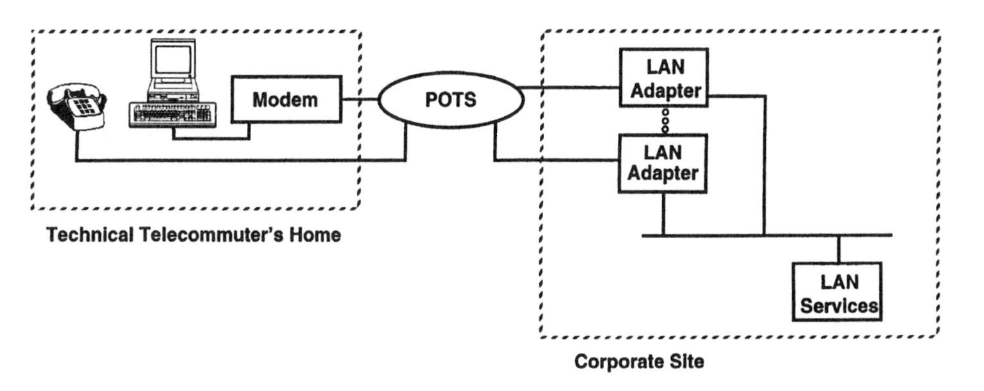

Figure 4.1 POTS-based technical telecommuter's links with corporate site.

office. This solution is popular because it is ubiquitous; it requires no changes to existing equipment, it supports voice equipment operation during local power outage, it is supported by telephone operations and well-established transport and interface standards, and it has a low equipment cost.

This solution, however, has several limitations. First, one POTS line operating at 28.8 Kbps (57.4 Kbps or higher with compression) may be inadequate to support the data bandwidth requirements of the technical telecommuters. Second, the technical telecommuters incurs the cost of two separate lines. Third, no standard fall back is available. Fourth, if a POTS line fails, it cannot be remotely isolated. Finally, modems could step down to lower speeds (e.g., from 28.8 Kbps to 19.2 Kbps or 9.6 Kbps).

4.4.2 The ISDN Solution

Three ISDN solutions that can partially address the technical telecommuter's networking requirements are examined next: the first solution meets the needs of the technical telecommuter for *high-speed data transfer and voice communications*; the second solution meets the needs of the technical telecommuter for *medium-speed file transfer, voice communications, and access to a frame relay network*; and the third solution meets the needs of the technical telecommuter for *data and voice conferencing*.

4.2.2.1 The High-Speed Data Transfer and Voice Solution

This solution, which is shown in Figure 4.2, provides the technical telecommuters with (2B + D) capabilities. The technical telecommuter can use two B channels for data, providing him or her with 128 Kbps of bandwidth. When the technical telecommuter receives a voice call, data transmission is reduced to one data channel, providing the technical telecommuter with 64 Kbps.

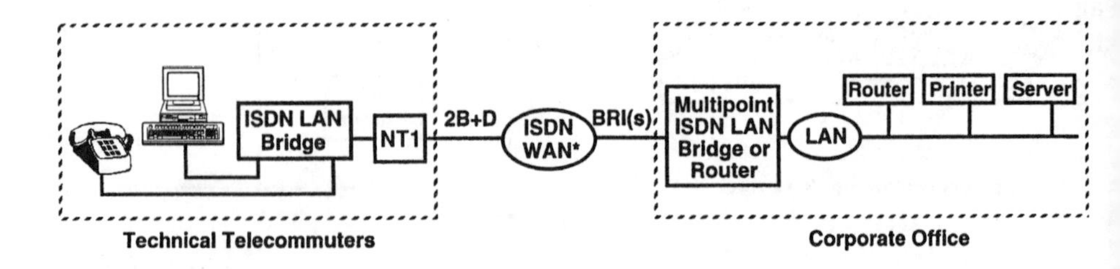

Figure 4.2 Technical telecommuter/corporate office link: the ISDN solution.

Through this solution, a technical telecommuter can access corporate e-mail, access corporate directories, make and receive phone calls, access the Internet and corporate LAN resources (e.g., printers), and transfer relatively large files.

As Figure 4.2 shows, to establish an ISDN link with the corporate office, a technical telecommuter needs [5] an analog phone and a personal computer or workstation equipped with applications compatible with LAN interfaces and the network operating system (NOS) located at the corporate site. This NOS routes traffic to and from the technical telecommuter's PC without noticing ISDN at the corporate end.

The technical telecommuter also needs an ISDN-ready LAN bridge, connected to both the workstation and the telephone, that has the ability to forward LAN datagrams to the multiport bridge or router located at the corporate site and to learn addresses dynamically. LAN datagrams are exchanged between these bridges by two simultaneous data calls, and provide the technical telecommuters with access to corporate LAN resources (e.g., printers, file servers, and computers). These bridges and routers also have security features and the ability to automatically hang up or dial, based on traffic. To perform these functions, both bridges must have (1) compatible functionality; (2) address screening capabilities based on a calling number or MAC address; (3) B-channel capabilities, including (if using NI-1) support for two directory numbers (DNs) and service profile ID (SPIDs), simultaneous data calling on two B channels, simultaneous voice and circuit data calling, and support for calling number identification services. These bridges must also have the capability to combine two 64-Kbps signals to/from the BRI into a 128-Kbps signal in a way that is compatible with the corporate end. In addition, the multiport bridge or router on the corporate site must have multiple BRI capabilities.

This solution offers the technical telecommuter several benefits over the dialup/modem solution, including: (1) higher bandwidth (128 Kbps of data bandwidth), meeting the file transfer space requirements of most technical telecommuters; (2) guaranteed delivery of information at 64 Kbps or 128 Kbps, depending on whether one or two B channels are used; (3) the ability to receive and make phone calls during data communications sessions; and (4) the ability to simultaneously handle voice and data on the same line.

This solution, however, does have the following limitations.

- ISDN requires local powering of all phones.
- It does not provide loop backup powering during a local power outage.
- Some rewiring of telecommuter premises will be needed to install ISDN.
- It has higher equipment costs than POTS. The price of a bridge or router could be a purchasing barrier if the technical telecommuter has to purchase this bridge instead of relying on the corporation. The price of bridges, however, continue to drop. (The price of the bridge and the NT1 are around $400.)

4.2.2.2 The ISDN/Frame Relay Solution

Some corporations have implemented or are considering the implementation of a frame relay network. At the beginning of 1994, there were 600 frame relay customers with about 500 installed ports. By 1997, according to Vertical Systems, there will be 5,500 companies using 65,000 frame relay ports. For those corporations, a hybrid frame relay/ISDN solution is emerging [6]. This solution provides the technical telecommuter with two B channels (Figure 4.3).

As Figure 4.3 shows, to establish an ISDN link with the private frame relay network the corporate office, technical telecommuters need a personal computer or workstation equipped with communications software, an analog phone [7], and a terminal adapter that is connected to both the PC and the analog phone. The adapter must have the ability to (1) assemble PC data into frame relay frames and disassemble frame relay frames into personal computer data; (2) support at least one frame relay permanent virtual circuit (PVC) on a semipermanent circuit-data B channel; (3) maintain one nailed-up circuit data connection trough ISDN to frame relay network, using one circuit-data B channel; (4) support circuit-data access at 56 Kbps or 64 Kbps over nailed-up connection; (5) support PVC and frame format in accordance with ANSI T1.618, *Core aspects of Frame Relay*, and Q.922, Annex A; and (6) support B channel capabilities, including voice calling on a single B channel, circuit data (semipermanent) on a single B channel, and two directory numbers (DNs) and SPIDs.

The technical telecommuter accesses the host through a frame relay network that must have one frame relay PVC between the NI-1 network interface and data interface serving host, in accordance with the same ANSI standards mentioned above. The host must have the capability to assemble host data into frame relay frames and vice versa.

This solution offers the technical telecommuter several benefits, including (1) guaranteed delivery of data at 64 Kbps, (2) the ability to receive and make phone calls during data communications sessions, (3) the ability to simultaneously handle voice and data on the same line, and (4) lower monthly

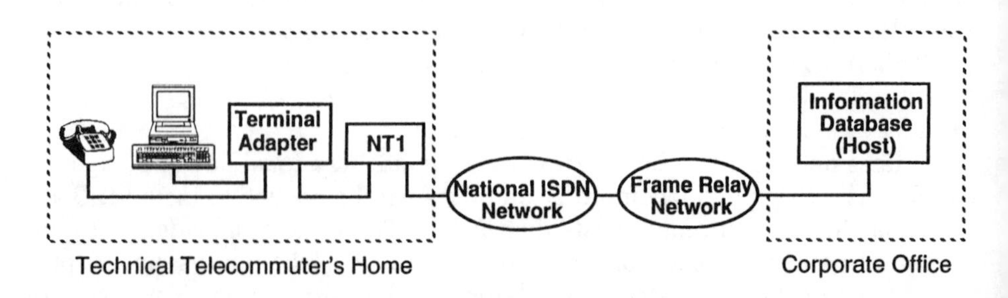

Figure 4.3 ISDN-based links of technical telecommuters with private corporate frame relay network.

cost than POTS. This ISDN solution also enables an IS manager to rely on existing frame relay infrastructure to establish links between the technical telecommuter and the corporate office.

This solution, however, does have the following shortcomings.

- The bandwidth offered by one B channel may be inadequate to support the requirements of the technical telecommuter.
- ISDN requires local powering of all phone equipment.
- It does not provide loop backup powering during a local power outage.
- Some rewiring of telecommuter premises will be needed to install ISDN.
- Higher equipment cost than POTS: The price of frame relay and related equipment is high. This could be a market barrier if the technical telecommuter has to purchase this adapter instead of relying on the corporation.

4.2.2.3 The Data- and Voice-Conferencing Solution

This solution, which is shown in Figure 4.4, provides the technical telecommuter with the ability to participate in a LAN-based screen-sharing conference that is taking place at a corporate location and to conduct a voice conference with the LAN participants as well. The voice conference is established using an existing non-ISDN conference bridge. The LAN-connected participants then establish screen sharing sessions with the LAN conference bridge. The LAN conference bridge supports multipoint conferencing functions. As a result, the software on each participant's PC only needs to handle a point-to-point conference with the bridge. The technical telecommuter establishes a circuit-data connection to the communication server which makes the computer appear to be on the LAN. The technical telecommuter then runs the same software as that used to set up a conference with the LAN conference bridge [8].

As Figure 4.4 shows, to establish an ISDN link with the corporate office, a technical telecommuter needs an analog phone that enables the technical

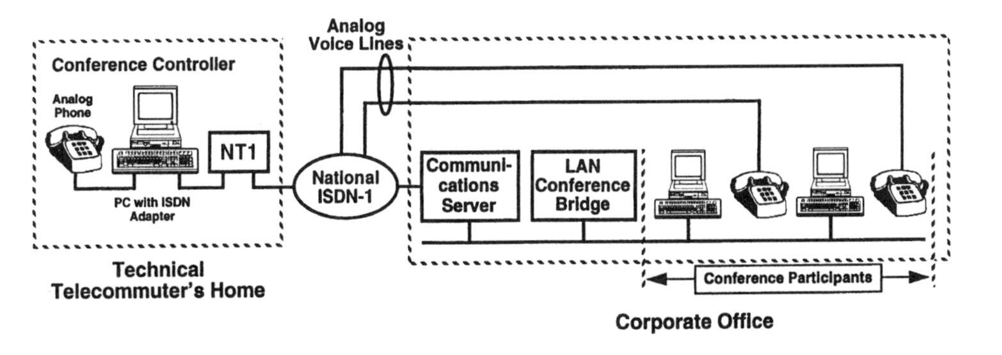

Figure 4.4 ISDN-based data conferencing links of technical telecommuters.

telecommuter, acting as a conference controller, to establish a voice conference using a non-ISDN conference bridge. The technical telecommuter also needs a PC or workstation equipped with an ISDN adapter and applications software compatible with software on corporate site. This PC also has several B-channel capabilities, including simultaneous voice and circuit-data calling, (if using NI-1), support for two directory numbers and service profile ID, and support for rate adaptation for rates lower than 64 Kbps.

On the corporate side, a communications server is required. This server must have application software compatible with communications software and with far end. This server must also have multiple BRI capabilities (the same B-channel capabilities as the PCs).

A LAN conference bridge and PCs are also needed at the corporate site. These bridges and PCs must have LAN Interfaces with application software compatible with LAN software, LAN devices, and LAN interfaces (e.g., Ethernet). In addition, the LAN conference bridge must have multipoint conferencing functions to support four or more users.

4.5 CONCLUSIONS

Some of the major applications that drive the communications requirements of technical telecommuters are voice communications, electronic mail, large-file transfer, remote access to databases, and data conferencing. Two major groups of networking solutions have been examined: dialup modem with two lines and three ISDN solutions.

Table 4.1 summarizes the findings of this chapter by comparing the strengths and limitations of the dialup modem and ISDN solutions.

Table 4.1
A Comparison of Technical Telecommuter Networking Solutions

User Selection Criteria	*Dialup Modem Solution*	*ISDN*
Applications supported:		
Large file transfer and voice		•
Voice access to frame relay		•
Data and voice conferencing		•
Small file transfer/voice	•	•

Table 4.1 (continued)
A Comparison of Technical Telecommuter Networking Solutions

User Selection Criteria	*Dialup Modem Solution*	*ISDN*
No rewiring required	•	
Low home-based equipment cost	•	
Single access line		•
Reliability:		
Guaranteed delivery of bandwidth		•
Support for voice during power outages	•	
Local powering of telephones	•	

As Table 4.1 shows, the dialup/modem solution addresses the requirements of the technical telecommuter for voice communications but does not meet his or her requirements for large-file transfer. However, the speed of modems has increased substantially over the years. ISDN solutions either meet the needs of the technical telecommuter for large-file transfer and voice communications or medium-bandwidth file transfer, voice communications, and access to a frame relay network, or they meet the needs of the technical telecommuter for data and voice conferencing.

References

[1] "Towards A National Research Network," National Research Network Review Committee, Computer Science and Technology Board, Commission on Physical Sciences, Mathematics and Resources, National Research Council, Washington, D.C.: National Academy Press, 1988, p. 15.

[2] Minoli, D., *Imaging in Corporate Environments*, McGraw-Hill, 1994.

[3] Sinnreich, Henry, and John F. Bottomley, "Any-to-Any Networking, Getting There from Here," *Data Communications*, Sept. 1992, p. 76.

[4] Quiat, B., "V.FAST, ISDN, or Switched 56," *Network Computing*, March 1, 1994, p. 70ff.

[5] "High-Performance Telecommuting (Interactive Graphics and Text)," NIUF Application Analysis Working Group, A Catalog of National ISDN Solutions for Selected NIUF Applications, Second Edition, Feb. 11, 1994, pp. 3-63 to 3-266.

[6] Tanzillo, Kevin, "ISDN-Frame Relay Link Is Good for Office And Remote Workforce," *Communications Week*, April 1993, p. 48.

[7] "Access to Frame-Relay Services," NIUF Application Analysis Working Group, *A Catalog of National ISDN Solutions for Selected NIUF Applications*, Second Edition, Feb. 11, 1994, pp. 3-135 to 3-138.

[8] "Screen Sharing with Workstations Connected to a Remote Local-Area Network," NIUF Application Analysis Working Group, *A Catalog of National ISDN Solutions for Selected NIUF Applications,* Second Edition, Feb. 11, 1994, pp. 3-33 to 3-35.

5 Mobile, Data, and Video Communications Applications for Marketing Telecommuters

Marketing telecommuters are highly mobile professionals with extensive needs for voice and data communications (as well as emerging needs for video communications) with the corporate office, their customers, and suppliers. This chapter explores the communications needs of the marketing telecommuter in order to provide the reader with specific examples of mobile, medium-bandwidth data and video communications applications.

The chapter introduces the reader to the functions performed by the marketing telecommuters and the value they create to their employers. The chapter then explores the communications links between the marketing telecommuters and their corporate offices, their communications and equipment requirements associated with the establishment of these links, and the potential networking solutions addressing these requirements. For each networking solution, some of the implementation details are provided and the strengths and limitations are described.

5.1 WHO ARE THE MARKETING TELECOMMUTERS?

The marketing telecommuter group is large, representing more than 21% of the corporate telecommuting population. This group performs a wide range of functions, including *sales, promotion, market research, planning,* and *pricing.*

Upper management of corporations is seeking to increase the value created by these marketing telecommuters by (1) maximizing the number of interactions between the marketing telecommuters and their customers to speed the process of identifying and selling new products and services, and by (2) improving the cost structure associated with the development and delivery of these products and services.

To accomplish corporate goals, marketers, whether they are working at home, in the office, or on the road, need to establish *cost-effective and rapid* networking links with other corporate employees, including their technical counterparts collaborating on the same project. They also need to establish communications links with other corporate functions (e.g., technical developers, inbound and outbound logistics). In addition, they need to establish communications links with their corporate suppliers (e.g., advertising agencies), and customers.

5.2 COMMUNICATIONS APPLICATIONS OF MARKETING TELECOMMUTERS

Most marketing telecommuters are highly mobile. They travel extensively and may spend two to three days a week away from the office either traveling or working at home. While they are away from the office, they need to duplicate to the extent possible the communications environment to which they are accustomed in the office. The communications applications of marketing telecommuters can be divided into three categories: (1) current voice applications, (2) current data applications, and (3) emerging video applications. Each of these categories of applications is described next.

5.2.1 Current Voice Applications

Voice communications are the most predominant communications needs of marketing telecommuters today. Voice applications requirements of marketing telecommuters are similar to the basic voice communications requirements of all the other telecommuters. Marketing telecommuters need to establish voice communications with the office to *access the corporate Centrex/PBX* system, enabling them to utilize voice communications features such as call handling, call hold, three-way calling, call transfer, consultation hold, call waiting, and calling number identification. They also need to access voice mail to retrieve messages from customers, supervisors, and suppliers. In addition, they need to have the messages that they receive from their customers transparently rerouted to their homes to overcome the reluctance of these customers to call them at their homes.

5.2.2 Current Data Applications

Five major data applications drive the data communications requirements of marketing telecommuters. Each is briefly described below.

1. *Electronic mail*: Marketing telecommuters need to exchange electronic messages and files with their colleagues and technical partners in the process of developing new proposals, contracts, and so forth. These collaborators may reside in corporate locations or in other locations.

2. *File transfer*: Marketing telecommuters need to download local files, such as advertising graphics, from the corporate office. They also need to retrieve corporate files, such as customer profiles and contracts. In addition, they may need to exchange files with their technical colleagues who are working on the same project.
3. *Access to databases*: Marketing telecommuters typically need to access a number of databases, including departmental databases, which may include information about specific accounts, markets, and products. They may also need to access public databases, which includes public information about the industry in which they operate. By accessing these databases, marketing telecommuters are in a better position to respond effectively to customer inquiries and/or to develop market forecasts while working at home.
4. *Data conferencing*: Marketing telecommuters may need to establish data conferencing sessions with their colleagues and supervisors located in the corporate office to: (1) jointly manage projects effectively; (2) hold new product discussions; (3) jointly develop documents; (4) exchange information about overall customer activity, sales leads and closure rates, and issues and problems of specific accounts; and (5) review and discuss marketing campaigns.
5. *Fax*: Marketing telecommuters need to send and receive fax-based documents from the office. For example, marketing telecommuters may need to fax sales proposals to the corporate office for management review and/or approval.

5.2.3 Videoconferencing Applications

Videoconferencing is a potential future communications application of marketing telecommuters. Whether they are working at home or in the office, these telecommuters may consider adding value to the visual dimension of their interactions with their trading partners. This would enable them to:

- Conduct face-to-face negotiations and transactions with customers;
- Conduct face-to-face general conversations for customer service;
- Develop advertising campaigns and review graphics with advertising agencies;
- Show/review products with suppliers and customers;
- View customer problems and demonstrate specific solutions to address these problems;
- Communicate with the corporate office while working at home.

By enabling marketing telecommuters to perform these functions, videoconferencing provides the marketing telecommuter with several benefits, including the reduction of long-distance travel, increasing the number of visual interactions with customers, and increasing the effectiveness of these interactions in comparison with audioconferencing.

5.2.3.1 Bandwidth Requirements of Marketing Telecommuters

Voice and data bandwidth requirements of most marketing telecommuters range from low to medium and could be satisfied with a 9.6-Kbps to 64-Kbps data connection. While most marketing telecommuters can be satisfied with the lower bandwidth range, market researchers and advertisers need the higher bandwidth in order to conduct extensive database searches or transfer files that include advertising graphics. Adding the visual component to the bandwidth requirements of marketing telecommuters increases their total bandwidth requirements to at least 128 Kbps.

5.2.3.2 Communications Equipment Requirements of Marketing Telecommuters

The communications equipment requirements of marketing telecommuters are shaped by their nature as mobile professionals. Marketing telecommuters prefer communications equipment that can perform multiple functions, and that are light, inexpensive, and fast. Two of the major equipment requirements of marketing telecommuters are (1) cellular phones; and (2) laptop PCs, preferably with built-in fax and modem capabilities. When the video communications application takes off, marketing telecommuters will also need either a videophone, or video conferencing capabilities incorporated in a PC.

5.3 NETWORKING SOLUTIONS

In establishing networking links between their homes and corporate offices, a number of networking solutions are available to marketing telecommuters. Three groups of communications solutions are explored in this chapter:

1. Data and voice communications solutions;
2. Videotelephony-based video and voice communications solutions;
3. PC-based data, and voice and video communications solutions.

For each type of communications solutions, at least two types of networking solutions are available: POTS and ISDN. Each of these solutions is explored next.

5.3.1 Data and Voice Communications Solutions

Voice and data communications solutions are the current predominant solutions that marketing telecommuters rely upon in establishing communications links with the corporate office and with their trading partners working for other corporations.

5.3.1.1 The Dialup/Modem Solution

As Figure 5.1 shows, to implement this solution, marketing telecommuters may rely on the following.

1. *A modem/fax operating at 9.6 Kbps–28.8 Kbps:* This modem/fax can either be directly attached to a PC or is an integral part of the PC. Combined fax/modems are becoming increasingly lighter and less expensive. For example, GVC Technologies offers a fax/modem with a speed of 14.4 Kbps. It is about the size of a deck of cards, runs on a 9-volt battery, and weighs 6.5 ounces. Prices of modem/fax combinations are priced approximately at $300 for DOS and Macintosh external units [1].
2. *A laptop, notebook, or palm-sized computer:* Some of the latest computers use personal computer memory card international association (PCMCIA) fax/modems to bring down the weight of fax/modems to mere ounces. For example, GVC offers a smaller version of its fax/modem, which can slip into the slot of PCMCIA-equipped notebook computers.
3. *Two POTS lines* that range in speeds from 9.6 Kbps to 28.8 Kbps. One of these lines is dedicated to voice transmission, providing the marketing telecommuter with access to Centrex services, while the other line is dedicated to data transmission.

At the corporate side, a pool of modems is dedicated to supporting the individual telecommuters. These modems are supported by a terminal server, providing the marketing telecommuter with access to LAN resources, such as e-mail, directory services, and corporate or departmental files.

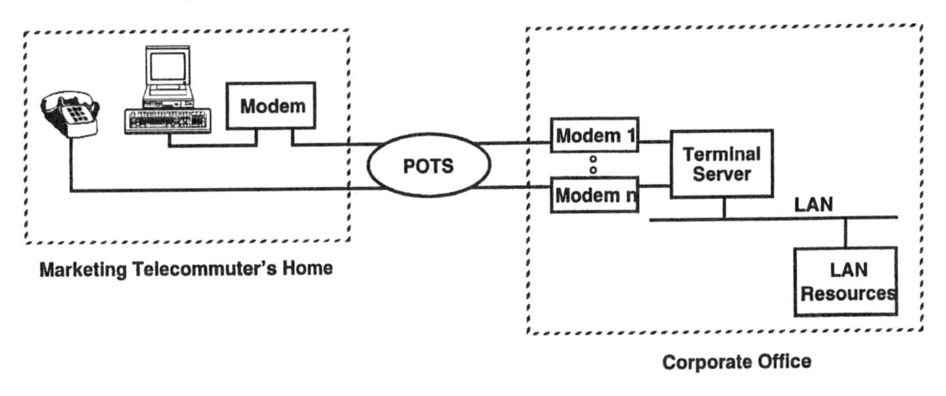

Figure 5.1 POTS-based voice and data communications links of marketing telecommuters.

The dialup solution is the present mode of operation (PMO). This solution is popular for several reasons: (1) it is ubiquitous; (2) it does not require any major changes to existing home-base equipment; (3) voice equipment operates during local power outages; (4) it meets the bandwidth requirements of most marketing telecommuters; (5) the cost of purchasing home-based equipment is low; and (6) transport and interface standards are well established.

This solution, however, is not optimal for several reasons: (1) POTS cannot simultaneously support the requirements of the marketing telecommuters for voice, data, and video communications; (2) the marketing telecommuters may incur the cost of two separate lines; (3) if a POTS line fails, it cannot be remotely isolated; and (4) the bandwidth requirements of marketing telecommuters, particularly advertising managers, may not be satisfied with a 9.6-Kbps to 64-Kbps data connection.

5.3.1.2 A Hybrid Cellular Phone/Dialup Modem Solution

To implement this solution, marketing telecommuters may rely on the following.

1. A cellular phone;
2. A modem/fax operating at 9.6 Kbps–14.4 Kbps;
3. A laptop, notebook, or palm-sized computer;
4. One POTS line that ranges in speeds from 9.6 Kbps to 28.8 Kbps. This line is dedicated to data transmission.

At the corporate side, a pool of modems is dedicated to supporting the individual telecommuters. In the previous case, these modems are supported by a terminal server, providing the marketing telecommuter with access to LAN resources, such as e-mail, directory services, and corporate and departmental files.

The hybrid dialup/cellular phone solution has several advantages: (1) it is a mobile solution, upon which a marketing telecommuter can rely whether at home or in a hotel; (2) it does not require any major changes to existing home-base equipment; (3) the marketing telecommuter only incurs the cost of one access line; (4) voice equipment operates during local power outages; (5) it meets the bandwidth requirements of most marketing telecommuters; (6) the cost of purchasing home-based equipment is low; and (7) transport and interface standards are well established.

This solution, however, is not optimal for several reasons: (1) it cannot simultaneously support the requirements of the marketing telecommuters for voice, data, and video communications; (2) if a POTS line fails, it cannot be remotely isolated; (3) the bandwidth requirements of marketing telecommuters, particularly advertising managers may not be satisfied with a 9.6-Kbps to 64-Kbps data connection; and (4) cellular service is still expensive—the user incurs the charge for both incoming and outgoing calls, including calls to 800 numbers.

5.3.1.3 The ISDN (2B +D) Solution

ISDN (2B + D) meets the needs of marketing telecommuters, market researchers, and advertisers for *voice communications* and *medium-usage data.* This solution enables the marketing telecommuter to use one B channel for voice and the other for data.

To implement this solution, the marketing telecommuter may rely upon (1) a digital ISDN telephone; (2) a desktop computer, a laptop, notebook, or palm-sized computer; and (3) an integrated terminal adapter. As Figure 5.2 shows, this adapter should be directly attached to the ISDN telephone, which in turn is attached to the personal computer. The adapter is also attached to a single access line, providing the marketing telecommuter with access to Centrex services and to corporate data communications resources.

At the corporate side, a pool of NT1s and terminal adapters (TAs) are dedicated to supporting the individual telecommuters. The number of NT1s and TAs represent half the number of modems associated with the basic dialup solution. As in the case of modems, NT1s and TAs are supported by a terminal server, providing the marketing telecommuter with access to LAN resources such as e-mail, directory services, and corporate or departmental files.

ISDN (2B + D) provides the marketing telecommuter with a total bandwidth capability of 128 Kbps in support of voice and data communications. If packet service over the D channel is supported by the carrier, then an additional 16 Kbps is available. This solution offers the marketing telecommuter several benefits over the basic POTS/modem line solution, including (1) higher bandwidth (128 Kbps of data and voice bandwidth), meeting the file transfer requirements of most marketing telecommuters; (2) higher reliability than POTS; (3) the ability to receive and make phone calls during data communications sessions; and (4) the ability to simultaneously handle voice and data on the same line.

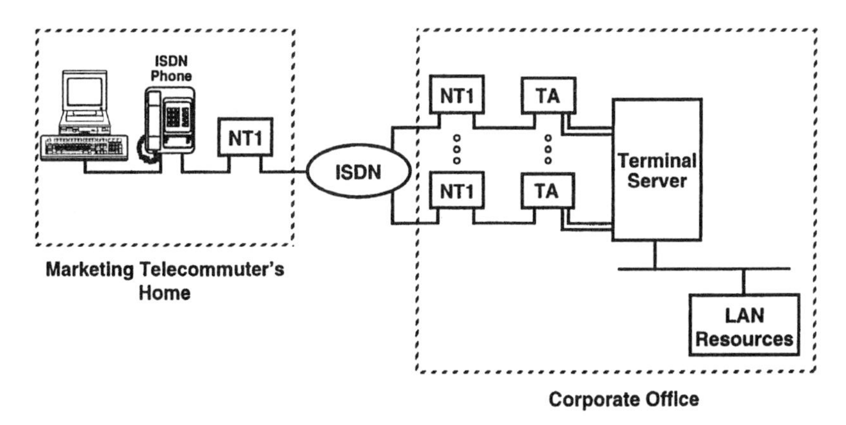

Figure 5.2 ISDN-based voice and data communications links of marketing telecommuters.

This solution, however, has the following limitations:

1. ISDN requires local powering of all phone equipment.
2. ISDN does not provide loop backup powering during a local power outage.
3. Some rewiring of telecommuter premises will be needed to install ISDN.
4. The price of an NT1 and an ISDN phone is not trivial. (The marketing telecommuter's cost may be $150–$300 for the NT1 and $200–$400 for the ISDN phone). This could be a market barrier if the marketing telecommuter has to purchase this equipment instead of relying on the corporation.
5. The overall cost of ISDN is generally higher than POTS, and the initial investment in ISDN equipment is more expensive.[1] As indicated earlier, the price of ISDN phone is in the $200–$400 range. This may represent a barrier to a marketing telecommuter who has to finance the purchase of this equipment out of his or her own pocket.

5.3.2 Videotelephony-Based Solutions

A number of videophones are re-emerging on the market to provide marketing telecommuters with the ability to establish video and audio communications sessions with their trading partners.

Videophones are not new equipment. They were introduced in the 1960s by AT&T. While the initial introduction of the videophone was not successful, renewed interest in video telephones has emerged in the last few years.

Currently, there are a number of videophones on the market. Some of these videophones can operate with POTS lines, while others operate with ISDN.

5.3.2.1 The POTS-Based Videotelephony Solution

Some videophones that are currently available on the market support the POTS solution. An example of these products is AT&T's 2500 videophone. This product supports modem rates of 19.2 Kbps, features a 3.3-in LCD, and has proprietary videocoding.

To operate a videophone, simply plug it into a standard POTS jack. Videophones are equipped with an internal modem to transmit audio and video signals over POTS facilities.

The POTS-based videotelephony solution offers the marketing telecommuter several benefits over the basic POTS solution, including (1) the ability to

1. For example, NYNEX charges $43 per month for an ISDN line. The usage is $0.05 per minute per B channel (if both channels are being used, the cost would be $0.10). The cost of a POTS line is $17 per month, and the usage is $0.02 per minute.

communicate with suppliers, customers, corporate colleagues, and supervisors via videoconferencing (with audio capabilities); (2) low cost of videophone equipment; and (3) plug-and-play capabilities.

This solution, however, has several limitations, including that (1) there is a lack of standards for video over the pubic switched telephone network (PSTN); (2) the POTS solution does not provide adequate bandwidth to support video and audio communications; (3) the video quality is closer to freeze frame; and (4) there is a limited number of suppliers offering videotelephones with POTS capabilities.

5.3.2.2 The ISDN-Based Videotelephony Solution

Several videophones are available on the market that support the ISDN solution. An example of these products is Northern Telecom's VISIT videophone. When first introduced, this product supported 56/64 Kbps, featured a 2-in by 4-in window on a Macintosh or PC, had proprietary videocoding with 16 gray levels, and required a TA to interface with ISDN BRI. Another example of an ISDN-based videophone is PictureTel's videophone, which is based on either proprietary standards or on H.261 standards and supports video from 56 Kbps to 128 Kbps. At the time of this book's writing, this videophone was manufactured by PictureTel in cooperation with Siemens.

As Figure 5.3 shows, the operation of ISDN-based videophones requires (1) an ISDN-based videophone at the marketing telecommuter's home as well as at the location of the person receiving the video call from the marketing telecommuter, and (2) an NT1 on both sides of the ISDN line [2]. As indicated above, ISDN videophones may operate at 128 Kbps, requiring two B channels or at 64 Kbps requiring one single B channel.

The implementation of a standardized ISDN-based videotelephony requires compliance with the following ITU-T standards.

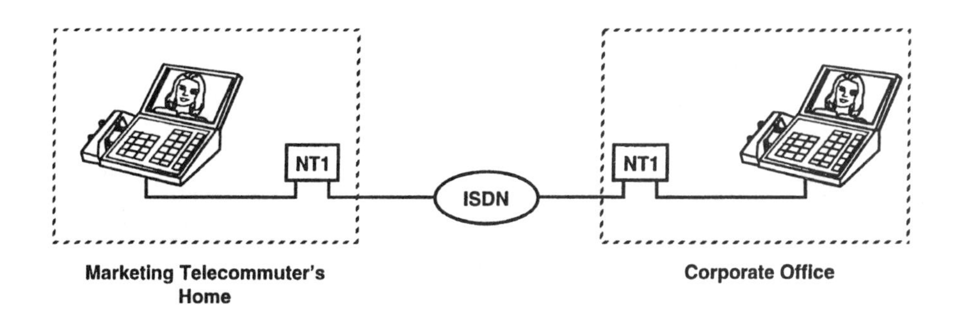

Figure 5.3 ISDN-based videotelephone links of marketing telecommuters.

- Recommendations H.2XX (e.g., H.221, H.230, H 242 and H.261) [3–5] focus on video coding, systems aspects for video communications, and high-resolution image transfer over ISDN at 64 Kbps and at multiples of 64 Kbps.
- Recommendation H.320 [11] defines compatibility of video terminals, including ISDN videophones, that operate using either one or two B-channel connections.
- Recommendations G.7XX (e.g., G.711, G.722, G.725, and G.728) [6–10] focus on the audio coding and systems aspects.

The ISDN-based videotelephone solution provides the marketing telecommuter with several benefits over the POTS-based videotelephony solution, including (1) a solution based on international standards; and (2) higher bandwidth than POTS (128 Kbps or 64 Kbps), resulting in higher audio and video quality.

The IDSN-based videotelephone solution has, however, several limitations, including (1) the fact that video calls based on national ISDN standards (NI-1, and NI-2) cannot reach POTS and other non-NI-ISDN users; (2) this solution provides limited feedback that may be useful to the marketing telecommuter, such as intercepts and other announcements; and (3) the audio quality of this solution is limited.

5.3.3 Voice and PC-Based Data and Videoconferencing Solution

Personal computers provide marketing telecommuters with an alternative to videophones in establishing video communications sessions with trading partners. Several pieces of equipment need to be assembled so that a PC can support the videoconferencing solution. These pieces include (1) An ISDN PC, (2) a microphone and speaker attached to each PC (or a standard telephone with a speaker phone), (3) a video camera, and (4) software to support a video window on a PC (Figure 5.4).

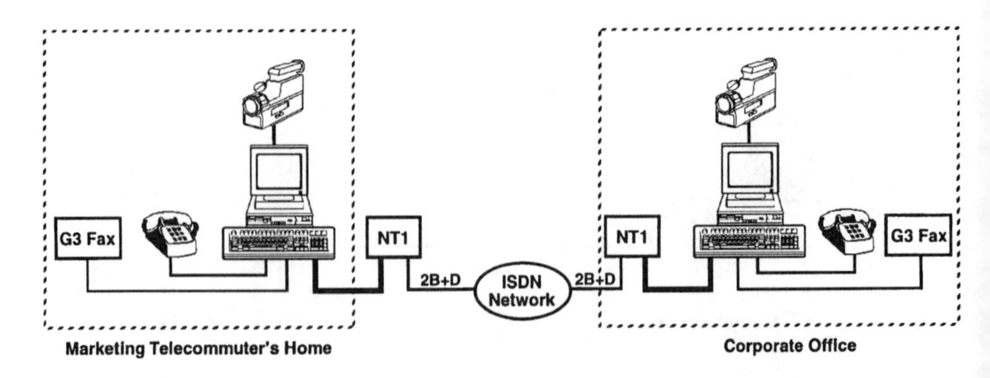

Figure 5.4 ISDN-based video, data, and voice conferencing links of marketing telecommuters.

To establish an ISDN-based videoconferencing session, the following steps are required:

1. Assuming the marketing telecommuter is the one originating the video call, he or she is designated as the conference originator, while the called party is designated as the conference recipient.
2. The marketing telecommuter places two circuit mode data (CMD) calls from his or her ISDN PC. These calls are transmitted by the ISDN network through two separate B channels. These channels are synchronized through the hardware on each PC.
3. The marketing telecommuter may also place an additional PMD call to establish a data communications link with his or her trading partner. This call is transmitted over the D channel of the ISDN line.

5.4 CONCLUSIONS

The communications requirements of marketing telecommuters are driven by their mobile nature and by basic voice applications, and low to medium data bandwidth applications. In addition, videoconferencing may become an important requirement of marketing telecommuters in the near future.

Several networking solutions are available to meet the communications requirements of the marketing telecommuter. Some of these solutions have been explored in this chapter, including the following.

- A number of data and voice solutions, including POTS with two lines, a hybrid cellular phone/dialup (with one line), hybrid cellular phone/ISDN (1B + D), and ISDN (2B + D).
- Two videotelephony-based solutions (ISDN and POTS).
- An ISDN-based solution supporting voice communications and PC-based medium-quality videoconferencing and data communications.

Table 5.1 (see next page) compares the strengths and limitations of the modem/dialup, the cellular phone/dialup, and the ISDN (2B + D) solutions.

Table 5.1
A Comparison of Marketing Telecommuter Networking Solutions

User Selection Criteria	*Dialup Modem*	*Cellular/ POTS*	*ISDN (2B + D)*
Applications supported:			
Medium-bandwidth file transfer + voice			•
Medium-quality video + voice conferencing			•
PC-based medium-quality video + data			•
Low-bandwidth data + voice	•	•	•
Mobility	•	•	
No rewiring required	•	•	
Low home-based equipment cost	•		
Single access line		•	•
Reliability:			
Guaranteed delivery of bandwidth			•
Support for voice during power outage	•	•	
Local powering of telephones	•	•	

References

[1] Scisco, Peter, *The Portable Office-Fax without Waiting*, 1993.

[2] "A Catalog of National ISDN Solutions for Selected NIUF Applications," Draft 5, Oct. 1992.

[3] ITU-T Recommendation H.221, "Frame Structure for a 64 Kbps Channel in Audiovisual Teleservices," Study Group XV, Report R37, Aug. 1990.

[4] ITU-T Recommendation H.230, "Frame Synchronous Control and Indication Signals for Audiovisual Systems," Study Group XV, Report R37, Aug. 1990.

[5] ITU-T Recommendation H.242, "System for Establishing Communication Between Audiovisual Terminals Using Digital channels up to 2 Mbps/s," Study Group XV, Report R37, Aug. 1990.

[6] ITU-T Recommendation H.261, "Video Codec for Audiovisual Services at p × 64 Kbps/s," Study Group XV, Report R37, Aug. 1990.

[7] ITU-T Blue Book, "Pulse Code Modulation (PCM) of Voice Frequencies," Vol. III.4, Recommendation G.711, 1988.

[8] ITU-T Blue Book, "7 kHz Audio Coding Within 64 Kbps/s," Vol. III.4, Recommendation G.722, 1988.

[9] ITU-T Blue Book, "System Aspects for the Use of the 7 kHz Audio Codec Within 64 Kbps/s," Vol. III.4, Recommendation G.725, 1988.

[10] ITU-T Blue Book, "Coding of Speech at 16 Kbps/s Using Low-Delay Code Excited Linear Prediction (LD-CELP)," Vol. III.4, Recommendation G.728, Study Group XV Report R 67, Dec. 1991.

[11] ITU-T Recommendation H.320, "Narrow-band Visual Telephone Systems and Terminal Equipment," Study Group XV Report R37, Aug. 1990.

Computer/Telephone Integration Applications for Home Agents

6

This chapter introduces home agents to the reader, and identifies their communications needs as well as those of call center managers and IS managers. The chapter then describes emerging computer telephone integration (CTI) solutions and associated networking solutions to meet the communications requirements of these groups. The chapter is concluded by identifying the risks and the benefits of CTI.

6.1 WHO ARE THE HOME AGENTS?

Home agents are individuals who operate out of their individual homes, are employed by a corporation either part-time or full time, and whose primary tasks are answering and/or placing calls on behalf of their employers. These home agents are managed by call centers, which typically operate as profit centers and are managed by a full-time manager. Call centers also manage those corporate agents who operate from central corporate call centers or from regional telecommuting centers (RTCs).

These agents may play one of the three following roles [1].

1. They may be dedicated to *outbound calling*. The agents in this case initiate calls on behalf of their corporations to existing or potential corporate customers. They place these calls either to introduce new products or to respond to customer inquiries regarding existing products. Examples of these agents include outbound telemarketers and surveyors.
2. They may be dedicated to *inbound call handling*. In this case, they respond to calls initiated by customers and routed to them through the corporate call center. Examples of these agents include inbound telemarketers, help-line operators, customer service representatives, and

order entry clerks. These employees have several responsibilities, such as providing their customers with product information and responding to customers' complaints.
3. They may be dedicated to *both incoming and outgoing calls.* In this case, they may perform some of the functions performed by agents who belong to the categories mentioned above.

According to industry estimates, the number of home agents was about 100,000 in 1992. While their number is small when compared to the number of telecommuters participating in other groups, their number is expected to grow steadily, particularly considering that the number of telemarketers as a whole is growing exponentially and, as will be explained later in this chapter, home agents can perform the same functions performed by the call center-based agents at a lower cost when given the right tools.

Corporations are seeking to maximize the value that these agents provide to the customers and to the corporation, both by increasing the revenues that these agents generate either directly or indirectly and by reducing the operating cost that the corporation incurs in hiring and training these agents. Call center managers need (or should seek) to translate these corporate goals into the following specific objectives.

1. Increase the number of calls received by or made by these agents.
2. Select technologies that maximize the value that these agents create.
3. Reduce the cost associated with (re)hiring and training these agents.

6.2 COMMUNICATIONS REQUIREMENTS OF CALL CENTER MANAGERS

Home agents play a minor role in selecting and implementing a home-based call center. Most of the decisions related to the establishment of the communications links of these remote agents are made by the call center managers. Call center managers have several technological requirements that would enable them to accomplish the objectives mentioned above. These requirements include the following.

1. *Easy access to management information systems (MIS)*: Corporations would like to provide their home agents with easy access to MIS. Better information lets call center managers track all parts of a business transaction, not just the calls, enabling them (the corporate agents) to manage themselves, to increase their motivation, and increase their job satisfaction [2].
2. *Synchronized access to voice and data*: Call center managers would like to provide home agents with simultaneous access to voice-based

customer calls, data-based customer information (e.g., account balance), and application environments.

3. *Voice messaging*: Call center managers now place a high priority on technological capabilities that can provide the caller with the option of leaving a message rather than waiting.
4. *Network-based information*: Call center managers need network capabilities that enable the home agents to identify information about the caller, including the caller's billing number, station number, and why he or she is calling. This information should be available to the agent by the time the caller is connected.
5. *Cost-effective call distribution*: A call center manager would like to cost effectively distribute incoming calls to the corporation among remote agents.
6. *Discriminatory queuing*: Call centers would like to be able to identify favored customers from call queues and give them special attention, automatically routing customers to their preferred agents instead of dumping these customers into a generic queue.
7. *A business case*: Call center managers sometimes need the support of vendors in developing business cases that can demonstrate to their upper management how the CTI benefits outweigh the cost.

6.3 NETWORKING REQUIREMENTS OF IS MANAGERS

In addition to the requirements of call center managers, IS managers have their own set of requirements. These requirements are primarily related to networking, and include the following.

1. *Computer telephone integration* [3]: Corporations may need CTI solutions, which refer to solutions that integrate their automated call distributors (ACDs) with the network, the host computer, other ACDs, and of course with the home agents.
2. *Expandability*: Corporations wish to buy ACDs that have the resources, processor capacity, and port capacity to incorporate additional clusters of corporate agents at remote locations.
3. *Standards*: Call center managers prefer standards that provide interoperability among the equipment offered by various vendors, providing them with the flexibility to choose among different vendors.
4. *A network migration strategy*: IS managers need to develop an information systems strategy that enables them to migrate smoothly, and with minimum risk from a small CTI system to a more advanced CTI system.
5. *Cost-effective access*: IS managers place a high priority on cost effective solutions to remote home agent access to the call center.

6. *Adequate Bandwidth*: IS managers need to provide home agents with adequate bandwidth to support their needs to access MIS databases, corporate or CO-based voice processing units. These needs could be met by a 9.6-Kbps to 28.8-Kbps connection.

6.4 COMPUTER TELEPHONE INTEGRATION SOLUTIONS

Several CTI solutions exist that address (most of) the requirements of call center managers mentioned above. These solutions are based on some or all of the following physical elements:

- A *switch*, which may be central office-based (Centrex), customer premise-based (PBX), or service bureau-based.
- A *computer system*, which may be based on a mainframe, a minicomputer, a microcomputer, or a departmental LAN.
- A *telephone network*, which may be ISDN-based or POTS-based.
- A *CTI server*, which is typically a separate/dedicated PC or workstation. In some cases, this server is a part of the computer system.
- A *voice processing unit (VPU)*, which is typically a separate voice response and/or a voice mail system. In some cases, the VPU is part of the mini or mainframe computer.

Each of the CTI physical elements performs specific functions. The switch (PBX or Centrex) performs several functions including switching, call answering, queuing, routing, and dialing. The computer system stores CTI applications and associated databases. These databases contain customer records, product lists, pricing information, inventory status, and so forth. The computer system may have a front-end processor (FEP), which reduces the capacity loads of the mainframe, decreasing potential mainframe upgrade cost. The FEP also provides fault tolerance and access to multiple databases on multiple computers. The various computer components (mainframes, minicomputers, and servers) are tied together via a local area network (LAN). The telephone network provides the agent with access to automatic number identifications (ANI), the calling line identification (CLID), and the dialed number identification service (DNIS) in addition to other advanced data services. The network also may tie together multiple switches or may move calls between sites. The VPU obtains information from and provides information to the customers (both calling and called parties) to the corporate agents (working at home or in the call center) and the call center supervisor. The VPU also gives callers the option of leaving messages rather than waiting. The CTI server establishes communications links between the PBX and the host computer, enabling them to exchange messages.

To perform the functions described above, each of the CTI physical elements incorporates several software and/or firmware elements. The telephone switch incorporates firmware, including ISDN network interfaces and CTI adapter boards, and may also incorporate existing switch processor upgrade/modification and, possibly, memory upgrade. The switch also incorporates CTI software, and may incorporate ACD software and networking software. The computer hardware incorporates CTI/data communications boards, and may also include memory and CPU upgrades, LAN/WAN upgrades, and a front-end processor. Computer software supports applications program interfaces, system management, and CTI enhancements to existing or new computer applications. Computer software may also include upgrades to operating systems, telecommunications management, database software upgrades, and predictive dialer software. The VPU incorporates firmware, such as telephone interface boards and voice boards. VPU software supports voice application, data communications, and voice application development, and may support computer data terminal emulation. The CTI server firmware includes switch interface/adapter boards and computer or LAN/WAN data communications boards. CTI server software supports switching to communication protocol conversion, system management, and data communications.

There are currently three CTI solutions that integrate some or all of these elements [4]: (1) a basic solution, (2) a semiautomated solution, and (3) a fully automated solution. Each of these solutions is described next. For each solution, the key CTI elements are identified and information flows are described.

6.4.1 The Basic CTI Solution

Figure 6.1 shows the traditional manual CTI solutions and the key elements associated with this solution. The information flows associated with this solution are as follows.

1. The call is routed to the home agent through the switch.
2. The home agent queries the calling or called person to determine his or her requirements.
3. The agent keys the customer's ID into the computer terminal.
4. As a result, information about the customers appear on the agent's screen.
5. The home agent then obtains the information that the caller requested from the computer database and provides the information to the caller or called person.

6.4.2 The Semiautomated Solution

Figure 6.2 shows the key elements of the semiautomated solution, which includes a VPU in addition to the other elements included in the traditional solution. The VPU is interconnected to the PBX and ACD as well to the computer database. This solution works as follows.

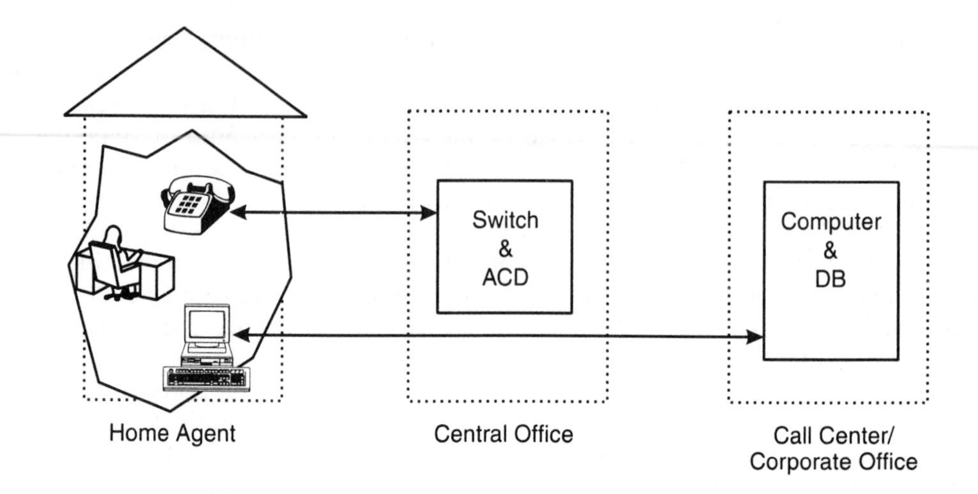

Figure 6.1 Home agent/corporate office link: the manual solution.

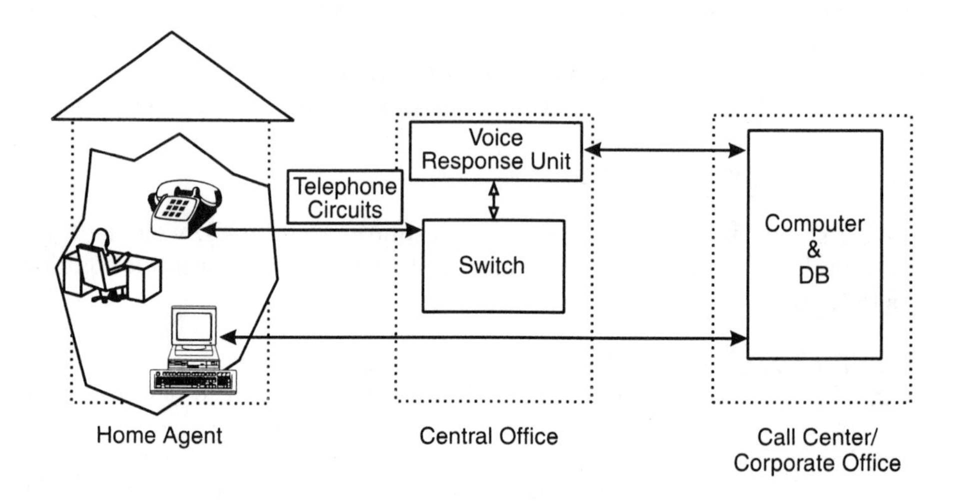

Figure 6.2 Home agent/corporate office link: the semiautomated solution.

1. The call is routed to the VPU initially through the PBX.
2. The VPU queries the calling or called person to determine his or her requirements.
3. The VPU then obtains the information that the caller requested from the computer and "speaks" the information to the caller or called person.
4. If the caller needs to speak to a live agent, the VPU initiates a transfer.

5. When the agent responds to the call, the VPU "whispers" the ID data to the agent.
6. The agent keys the customer's ID into the computer terminal. As a result, information about the customers appear on the agent's screen.

6.4.3 The Fully Automated Solution

The solution shown in Figure 6.3 is the most advanced solution available to date. This solution incorporates all the CTI elements. In addition, the PBX (or Centrex) is interconnected to the computer through a CTI link. This solution operates as follows.

1. The VPU queries the customer to obtain the required information.
2. The computer application synchronizes information received about the caller with available database information and automatically fills the screen with this information when the live agent receives the information.

A more detailed example of the networking aspects of this solution are provided in Section 7.7.2.

6.5 CTI NETWORKING SOLUTIONS

A home agent may be able to access a corporate CTI system by relying on one of two solutions: (1) the dialup/modem solution, and (2) ISDN. Each of these solutions is described next.

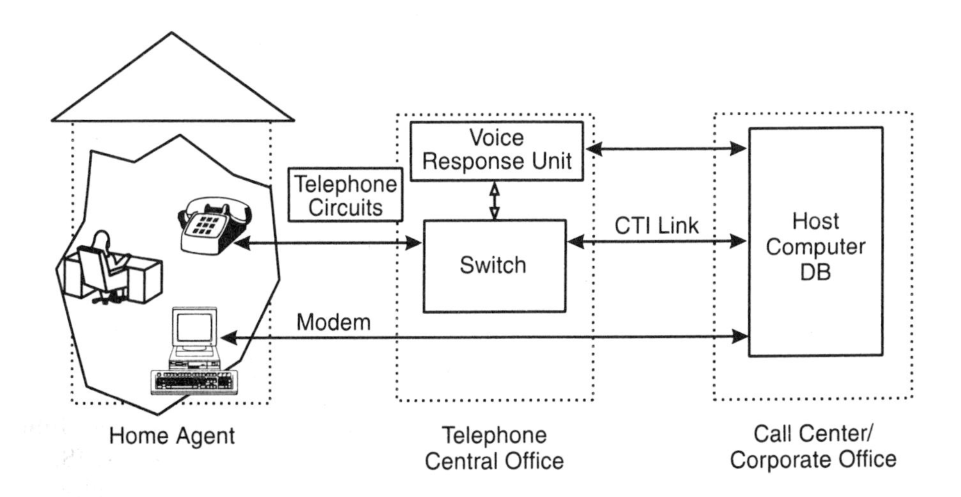

Figure 6.3 Home agent/corporate office link: the fully automated solution.

6.5.1 The POTS Solution

The dialup/modem solution is the predominant mode of operation (PMO) today. To implement this solution, home agents rely on the following equipment and services: an analog phone and a data terminal connected to a public telephone network. The home agent shares the POTS line for voice and data applications. This solution, however, is not ideal for several reasons: depending upon the amount of information that the home agents needs to access, the home agent may need two separate lines; no standard fallback is available; and if a POTS line fails, it cannot be remotely isolated.

6.5.2 The ISDN Solutions

Two ISDN solutions could address the needs of the home agent: ISDN (1B + D) and ISDN (2 B + D). As in the case of the administrative telecommuter described in Chapter 3, ISDN (1B + D) addresses the requirements of home agents for *voice communications and light data transmission*. The home agent uses the B channel for voice, while packet data is transmitted on the D channel. To establish an ISDN link with the corporate office, home agents need a digital telephone and a personal computer [5], which are linked to the ISDN network through an NT-1. ISDN B-channel capabilities include support for one directory number and service profile ID. The B channel also supports the home agent with multiple voice capabilities, including flexible calling (with six-way conference calling, hold transfer, and drop) as well as calling number identification services.

ISDN (2B + D) is a superior solution, providing the home agent with *voice communications and medium data transmission*. This solution is particularly useful to the home agent if he or she has more extensive data requirements (e.g., more extensive databases) than those that can be met with the ISDN (1B + D) solution.

As Figure 6.4 shows, this solution works as follows. A call that is received, for example, through an 800 number is terminated at the PBX. As the call passes through the ACD (which in many cases is an integral component of the PBX), the calling line ID (CLID) is broken off and transmitted to the host database, accessing the caller's records. As the phone rings at the agent's house, the home agents respond to the call via one of the two ISDN B channels, while the information retrieved from the host is received by the agent on the other B channel and appears on the PC screen. Throughout the call, the D channel is managing and controlling the call. It sets up both the voice and data channels and it is the pipeline used to collect call accounting information. When warranted, it can be used to coordinate screen sharing with the call center.

ISDN offers the home agent several benefits over the dialup/modem line solution, including a single access line and screen sharing capabilities. ISDN also offers the home agent higher reliability because it guarantees the delivery of information at the bandwidth promised to the telecommuter.

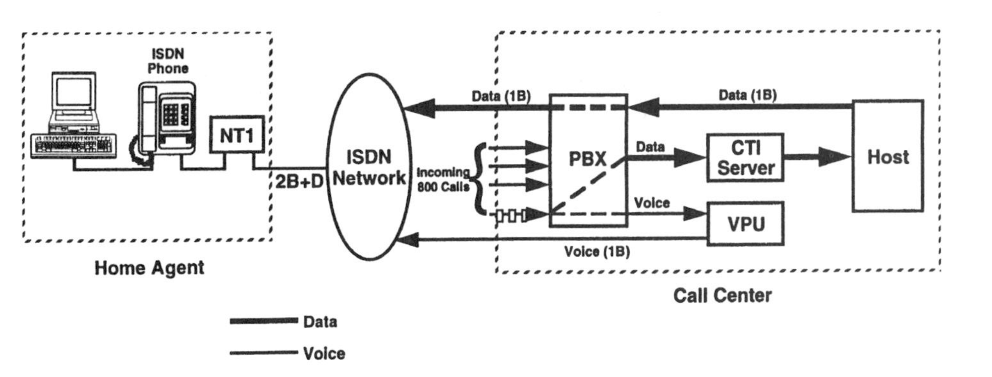

Figure 6.4 ISDN-based home agent's links.

ISDN, however, requires special, fairly expensive equipment (the ISDN phone), local powering of all phone and data equipment, some rewiring of the home agents' premises, and does not provide loop backup powering during a local power outage.

6.6 CTI COST

The call center manager who plans to implement a CTI system will incur several types of costs in certain areas, while reducing costs in other areas. The key cost components of a CTI solution include labor costs, equipment costs, and telephone access charges. Each of these cost components is described below.

Labor costs include the salaries, overtime, commissions, (re)training, and recruiting costs of the corporate agents. By hiring home agents, the corporate labor cost is likely to be reduced because the satisfaction of the home agents will result in the reduction of turnover rate. Further cost reductions can be accomplished if the home agents are only hired part-time. For example, these home agents could be hired during peak hours, during peak seasons (e.g., Christmas), in case of emergencies (e.g., the Tylenol scare, stock market crash, etc.), and during periods of peak demand (during a sale on certain merchandise).

Equipment/software-related costs will depend on the CTI solution that the IS manager selects and may include the initial cost of designing, selecting, acquiring, upgrading, developing, and installing some or all of the following equipment: ACD/PBX, audio terminal, VRUs, and data. Equipment cost also encompasses the operating cost associated with this equipment, including the cost of maintenance, training, and managing the equipment.

Facilities additions and upgrades includes the cost of leasing, heat, light, and so forth. By relying on home agents, this cost would be partially eliminated.

Telephone access charges includes the cost of an 800 service, an ISDN line, and local services. These charges should be lower than those incurred by the corporation prior to deploying a CTI system. These cost savings will be driven by the effect of reduced hold times.

Company-Specific Cost Factors

The total cost associated with the implementation of a new CTI system or the upgrade of an existing system can vary widely from one corporation to another and depends on several factors, including the current installed base of computers; CTI systems and so forth; the capacity available on existing PBXs and computers; the current level of expertise available on the computer side as well as the communications side of the IS organization; the type of computer solution selected, which may be based on a mainframe or a minicomputer; the type of system architecture selected, which may be centralized or decentralized; and the number of agents (both those located in call center and those working from their homes).

6.7 THE BENEFITS AND RISKS OF A CTI SYSTEM

As mentioned above, a CTI system combined with employing either part-time or full-time home agents can contribute to cost savings. Another obvious benefit of deploying a CTI system is revenue increase. These revenues would be generated by increasing the number of responses to calls, resulting in more customers.

Call center managers need to weigh these benefits of a CTI solution not only against the cost but also against the risks of deploying these systems. The sources of risks are significant.

First, CTI requires an information systems (IS) team that is knowledgeable in multiple disciplines, including voice-oriented technologies (PBX and VPUs) and data-oriented technologies (computers, LANs, etc.). Members of this team need to work closely to ensure that the CTI development and implementation is successful. This kind of cooperation may be difficult in corporations that have a clearly fragmented IS organization with a clear separation of voice- and data-related functions.

Second, the integration of multiple components is technically difficult and, depending upon the solution selected, may require the involvement of an external systems integration company.

Third, there are a various products and services offered by multiple vendors from a variety of backgrounds, including computer vendors, PBX vendors, and software vendors. The selection of the appropriate products is not an easy task.

Fourth, currently there are no standards supporting switch/computer communication protocol. Several protocols are emerging to support these protocols, including adjunct/switch applications interface (ASAI), which is supported by AT&T; and the switch to the computer applications interface (SCAI), which is supported by a number of vendors, including Northern Telecom.

6.8 CONCLUSIONS

The decision to implement a CTI system should be based on a sound business case that clearly measures the benefits of deploying CTI systems versus the cost of implementation. This chapter demonstrated that corporations can gain several benefits and may also incur extensive costs from deploying a CTI system. To ensure the success of a CTI system, and a home agent program, corporations need to consider the following factors:

1. Employing home agents can offer the corporations several cost benefits.
2. The planning stage is a crucial stage in the development of a successful home agent program. All the players involved from the IS organization(s) need to be involved from the start.
3. The project has to be led by an experienced project manager with extensive contacts throughout the organization.
4. The support of upper management is crucial.
5. The involvement of a reputable vendor at an early stage of a CTI project can also help maximize its chances of success.
6. A phased approach to the implementation of a CTI system is recommended. For example, those corporations that have no prior experience in CTI may want to consider the basic manual solution described in this chapter as the starting point.

References

[1] "Computer/Telephone Integrated Systems," *Vanguard Communications*, pp. 43–45.

[2] Quinn, Brian, "ACD Issues—Things to Keep in Mind When You Shop," *Teleconnect*, April 1993, pp. 98–101.

[3] "ACD Technology that Meets Your Needs," *Telemarketing*, Feb. 1993, pp. 46–47.

[4] "Computer/Telephone Integrated Systems," *Vanguard Communications*, pp. 108–112.

[5] "ACD Agent At Home," NIUF Application Analysis Working Group, *A Catalog of National ISDN Solutions for Selected NIUF Applications*, Second Edition, Feb. 11, 1994, pp. 3-51 to 3-52.

7 Private and Public Application-Level Solutions

In the last three chapters, a number of data applications were described, such as electronic mail, data conferencing, access to databases, videoconferencing, and file transfer. A number of networking solutions were also proposed to meet the networking requirements of the telecommuters associated with these applications. This chapter extends the identification and analysis of communications solutions by focusing on higher level application solutions. These are solutions that operate at layer 7 (or above) of the OSI reference model. Three application solutions are explored in this chapter:

1. Groupware;
2. The Internet solutions;
3. Value-added network solutions.

While groupware solutions are predominantly implemented by corporations as a private intracompany communications solutions, the Internet and VAN solutions are offered by public service providers. This chapter identifies the criteria that telecommuters and IS managers (should) take into account in selecting these application solutions. The chapter then provides a detailed description of the three solutions. For each solution, technical characteristics are described, benefits are outlined, the range of supported applications are defined, the key vendors or service providers are identified, and the strengths and limitations are highlighted.

7.1 IS MANAGEMENT REQUIREMENTS

Information system managers have a wide a range of criteria in selecting application solutions, which includes the following.

1. *Geographic coverage*: IS managers seek application solutions that support the geographic range of their end users. This range may encompass local area networks, metropolitan area networks, and wide area networks.
2. *Range of applications supported*: In comparing private solutions with public applications solutions, IS managers compare the range of applications solutions offered by the enhanced service provider (VAN providers or Internet providers) with the applications that a company can put together on its own (e.g., groupware). IS managers are concerned about a wider range of applications than those that are important to telecommuters. In addition to e-mail, database retrieval, and data conferencing (the key requirements of the telecommuters), IS managers may also consider other application solutions, including transaction processing, network management services, router network services, and EDI.
3. *Support for installed base*: IS managers value solutions that can integrate well with the current and future installed base of hardware, operating environments, messaging transports, and database management systems.
4. *Access methods and speeds*: IS managers have needs for different access methods and speeds in support of applications solutions. These requirements are shaped by their installed base as well as by cost factors. The range of access methods required by IS managers in support of application solutions may differ from one company to another. Access mechanisms required may include dialup, switched digital, dedicated digital, and dedicated analog.
5. *Integrated solutions*: IS managers prefer to reduce the cost associated with managing various types of messages (e.g., e-mail and fax) by either relying on a single mailbox to retrieve these messages or by relying on messaging gateways.
6. *Ease of use and administration*: IS managers value applications solutions that are easy to administer and to use by telecommuters and by other corporate end users.
7. *Low cost*: IS managers (should) take into account a number of elements in measuring the total cost associated with the implementation of a new application solution, including the cost of acquisition, implementation, maintenance, and training.
8. *Intracompany versus intercompany communications*: The appropriateness of the application solution may depend on the extent of the needs of IS managers for intercompany versus intracompany communications.
9. *Collaboration media* [1]: Applications solutions selected by IS managers need to support the information exchange media required by the end users (including those of telecommuters), such as multiuser database and file access, store-and-forward messaging, real-time messaging, and real-time computer conferencing.

10. *Outsourcing*: The appeal of a private versus a public application solution will be driven by the size of the corporation and whether IS managers have the resources to install, maintain, and upgrade an application solution or whether they prefer to outsource these functions to an external vendor.

A survey by Focus Data, Inc., shed light on the degree of importance of customer selection criteria. This information is shown in Table 7.1.

Table 7.1
Internet Access Service Selection Criteria Based on Highest Possible Score of 10.0

Criteria	*Importance Rating*
Ease of use	8.7
Price	7.9
Range of applications and services supported	7.8
Range of access methods available	7.7
Access protocol	7.5
Geographic scope of provider's backbone network	7.3
Number of gateways on provider's network	6.9
Training	6.2
Availability of direct international links	6.2

Source: Focus Data, Inc., Framingham

7.2 GROUPWARE SOLUTIONS

The term *groupware* refers to software that supports at least one of the following applications:

- Electronic messaging;
- Data conferencing;
- Process management;
- Messaging gateways.

Messaging is one the key applications of groupware. Groupware messaging applications can be used by telecommuters to perform a number of functions, such as communicating with other corporate employees, routing documents and electronic forms, and requesting schedules. Telecommuters can have access to these capabilities from their homes through dialup access.

Process management applications include time management, workflow management, and project management. These applications assist corporations in automating the flow of work between individuals and their workstations. For example, administrative telecommuters performing financial functions can rely on this application. Groupware routing databases establish a path for documents, enabling an administrative telecommuter to create a financial document, send it to as many people as necessary for approval, and monitor its progress [2].

Groupware computer conferencing applications enable telecommuters to interact with other employees in the office in concurrent collaborative ways. Through these capabilities, users can view, modify, and approve spreadsheets and other documents by viewing windows containing a *slate* or *whiteboard* on their individual personal computers and making individual contributions as though they were using different colored *pens.* For example, marketing telecommuters working at home, in the office, or in the field can use groupware databases to track overall customer activity, monitor leads and closure rates, and cooperate in resolving problems of specific accounts. Marketing and technical telecommuters can hold new product discussions through a groupware database. This database offers the technical and marketing collaborators a forum to share comments and opinions.

Messaging gateways provide corporate employees working at home or in the office access to public databases, private databases, fax, paging, and voice mail.

Emerging groupware videoconferencing applications: Video conferencing is a groupware application that is expected to emerge in the 1994–1995 timeframe. Through a small window on his or her desktop workstation, a telecommuter would be able to establish a videoconferencing session with other corporate collaborators [3].

7.2.1 Groupware Products

An increasing number of vendors offer groupware. The products offered by these vendors can be classified into three categories, as follows.

1. *Low-end groupware that supports a single application,* such as computer conferencing, project management, workflow management, and time management. Examples of these products are those offered by Fujitsu (DeskTop Conferencing) and ON Technology Corp. (Meeting Maker 1.5).
2. *Groupware that supports a single local area or workgroup,* but may support a number of functions. Examples of such groupware products include DaVinci's Systems.
3. *High-end software that supports multiple functions over wide area enterprise networks.* The three key products offering multifunctional

groupware are Lotus Notes, WordPerfect Office 4.0, and Digital Equipment Team Links.

Table 7.2 provides a description of the major three multifunctional groupware. As Table 7.2 shows, these groupware packages differ in terms of the number of platforms they support, the number of NOS they support, network architecture, and the range of applications they support [1].

Table 7.2
Multifunctional Groupware

Product Characteristics	*Team Links*	*Lotus Notes 3.0*	*WordPerfect Office 4.0*
Platform			
DOS	•	•	•
Macintosh	•	•	•
OS/2		•	•
Unix		•	
Windows	•	•	•
NOS Support			
LAN Manager		•	•
Netware	•	•	•
LAN server		•	•
Vines		•	•
Network Transport Support			
DECNET	•	•	
IPX/SPX		•	
NETBIOS		•	•
Named Pipes			
SNA		•	
TCP/IP	•	•	

Table 7.2 (continued)
Multifunctional Groupware

Product Characteristics	*Team Links*	*Lotus Notes 3.0*	*WordPerfect Office 4.0*
Messaging Gateways			
Other private e-mail	•	•	
Public e-mail	•	•	•
Fax	•	•	•
Paging		•	
Voice mail	•		
Conferencing applications			
Bulletin board	•	•	•
White board	•		
Authoring/editing	•	•	
Brainstorming	•	•	
Process Management Applications			
Time	•		
Workflow	•	•	•
Project/task			•
Price	$440–550 per server	$495 per user	$495 per server

Source: Network World

7.2.2 Access Methods Supporting Groupware

Most groupware applications can be supported by dial-in access and leased/switched 56 Kbps. The two exceptions are interactive concurrent collaboration and desktop video conferencing. These are more demanding applications that will require higher speeds.

7.2.3 Groupware Strengths and Limitations

Groupware offers telecommuters and other corporate employees several benefits, including the following.

1. *Low cost*: In implementing groupware, IS managers incur a nonrecurring cost. As Table 7.1 illustrates, this cost is either on a user or server basis.
2. *Multiple applications*: Groupware supports multiple applications, including computer conferencing, electronic messaging, business process management, and messaging gateways.
3. *Ease of use*: Multifunctional groupware solutions are based on graphical user interfaces, which makes it easy to use by telecommuters and other corporate employers. Consequently, groupware solutions are appropriate for all groups of telecommuters.
4. *Security*: As a private solution, groupware provides a secure solution for intracompany communications.
5. *Training and consulting*: The leading providers of multifunctional groupware provide their buyers with training and/or consulting as part of their product package.

While groupware offer corporations several benefits, IS managers need to consider that groupware is predominantly implemented as a private application solution. They also need to consider the cost of selecting, implementing, and upgrading the groupware solution.

7.3 THE INTERNET

The Internet is a networking and application solution that is becoming increasingly popular among technical developers and telecommuters. Several vertical industries are currently strong supporters of the Internet, including higher education, government agencies, and major corporations. The Internet is also becoming increasingly popular among healthcare providers, k-12 schools, and community colleges.

The Internet is an international network that supports over 34,000 computer networks and over 2 million computers worldwide. Estimates of the number of Internet users vary from 7 million to 12 million; half the Internet users reside in the United States. The Internet has been growing rapidly in the last several years. Several statistics reflect this fact: (1) the number of subscribers has been growing at 100% per year for the last three years; (2) the number of universities and research laboratories using the Internet has increased from 200 in 1988 to more than 10,000 in 1993 [4]; and (3) the number of packets carried on the Internet has increased from millions per day to billions per day.

7.3.1 The Internet Infrastructure

The Internet is a network infrastructure that supports research and engineering functions and is sponsored by a variety of federal agencies, including the National Science foundation (NSF) and the Advanced Research Project Agency (ARPA). The Internet is made up of three tiers of networks: a T1/T3 backbone (Figure 7.1), regional networks, and local networks [5].

The national backbone connects NSF-funded supercomputer centers throughout the country. The regional networks are midlevel networks centered around the NSF supercomputer sites. These regional networks support local or campus networks. In addition to these three tiers, a number of distinct commercial networks (e.g., CompuServe), have established gateways or bridges to the Internet.

7.3.2 Logical Internet Connections

The Internet logical connections are predominantly based on a suite of TCP/IP protocols. The IP layer is implemented on end systems as well as intermediary nodes. It is responsible for internetworking multiple networks with different underlying protocols. These underlying protocols may support a specific LAN (e.g., Ethernet/IEEE 802.3), metropolitan area networks (e.g., fiber distributed data interface), or wide area networks. Each IP node (e.g., a router or an end system) has a unique 32-bit address that enables it to identify and communicate with other nodes.

Transmission control protocol (TCP) and user datagram protocol (UDP) are transport protocols that are implemented on end points to packetize all application-layer exchanges and to provide transport services suitable for the applications. The transport layer controls the size of the packets, the number of packets that comprise an entire transmission, and the format of the data in the packet.

Several applications are supported by the transport layer. Some of these applications, such as simple mail transport protocol (SMTP), Telnet, and file transfer protocol (FTP) are supported by the TCP protocol while other applications, such as the simple network management protocol (SNMP) are supported by UDP (Figure 7.2). These applications drive many of the enhanced services offered by the Internet.

7.3.3 Internet Service Providers

Internet services are offered by a number of providers. These providers can be categorized into two major groups: national Internet providers, and regional Internet providers [5].

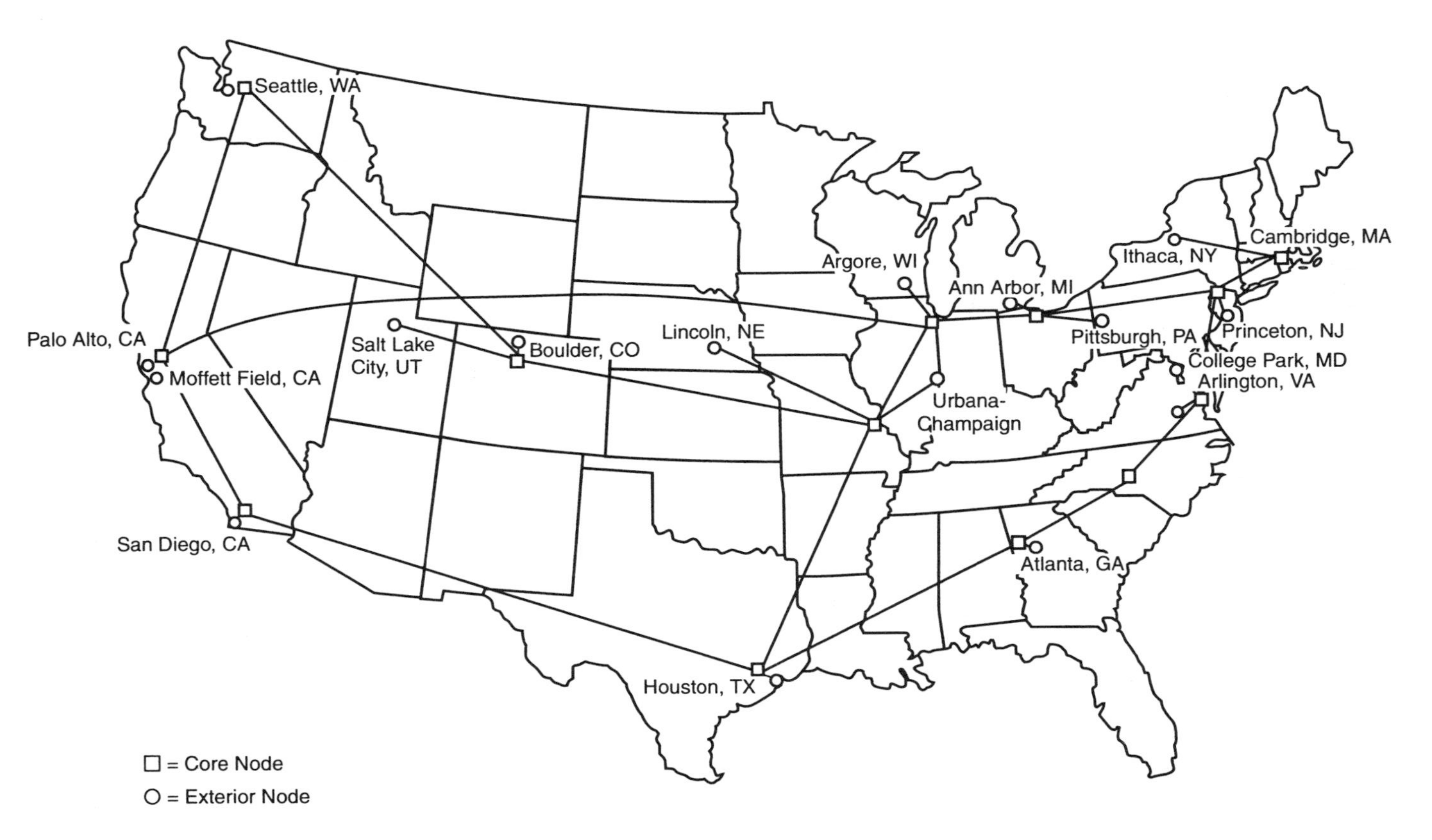

Figure 7.1 NFSNET T-3 backbone as of autumn 1992. (*After:* [5].)

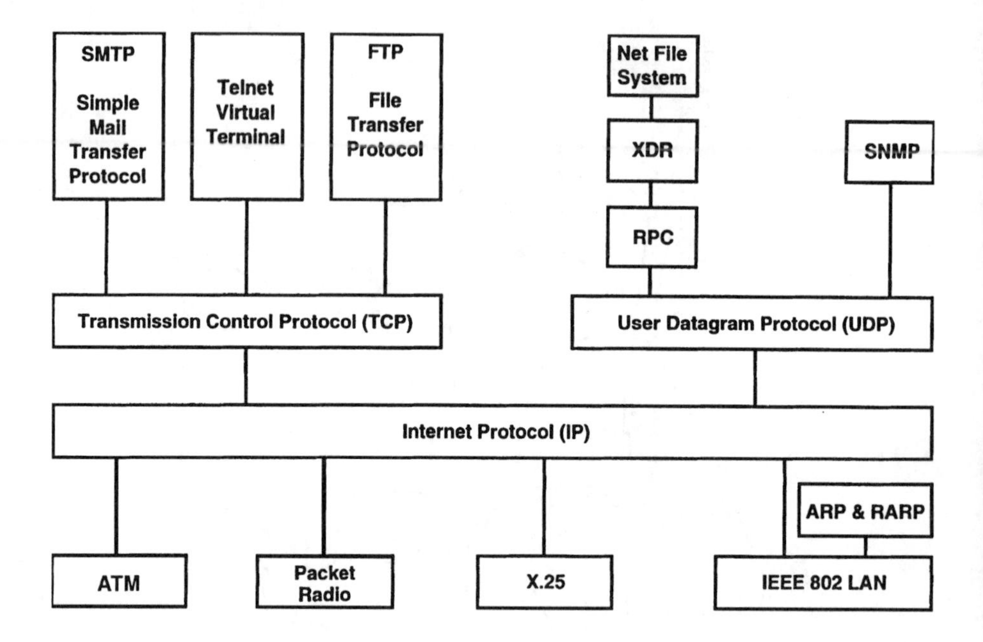

Figure 7.2 TCP/IP.

Each of these groups of providers differs in terms of their geographic presence, the number of services that they support, and the network solutions that they offer. Differences also exist among the providers that belong to the same group.

National Internet service providers operate in multiple cities across the United States. These providers provide participating members with access (gateways) to other domestic and international networks. They include the Advanced Network and Services, Inc. (ANS), California Education and Research Federation Network (Cerfnet), Performance Systems International, UUNet Technologies, Inc., and Sprint Corporation.

Regional Internet providers include the San Francisco Bay Area Research Network (BARRNet), the Committee on Institutional Cooperation Network (CICNet), and Southwestern States Network (WestNet), which operate in the western region of the United States. Global Enterprise Services Inc./Northeast Regional Network (JvNCnet), and New England Academic and Research Network (NEARNET) operate in the Northeast. CICNet also operates in the Midwest, while Southeastern Universities Research Association Network (SURANet) operates in the Southeast (Figure 7.3).

Figure 7.3 Internet regional access providers' coverage. (*After:* [5].)

7.3.4 Internet Applications

By accessing the Internet, the telecommuter can receive some or all of the following services:

- Logon services;
- E-mail;
- File transfer;
- Host-to-host communications;
- Directory services.

Each of these services is described next [4].

Logon services are offered by Internet providers through two types of TCP/IP-based protocols: *Telnet and rlogin.* The Telnet command has the advantage of being available on any host that allows remote login capability. Rlogin, a "Berkeley r command," on the other hand, is not necessarily available on any host that allows remote login capability. The advantage of rlogin, however, is that it passes on more information about the environment of the calling machine to the target host than Telnet does. Telnet is supported by most Internet service providers. Exceptions include Sprint, UUNet, and WestNet.

File transfer services offered by the Internet are based on a TCP/IP protocol, called the file transfer protocol, which is specified in RFC959 and is used to transfer files between Internet hosts. The technical telecommuter can use FTP to download as well as upload files. The FTP protocol supports multiple file formats, such as ASCII and EBCDIC.

Anonymous FTP is a simple and powerful extension to the basic FTP protocol, which enables a telecommuter to create and share public archives with others on the Internet. Currently, an extensive amount of information is available to Internet participants through anonymous FTP, such as public domain or freely available software, abstracts and full papers published by many researchers, public domain books, and Internet standards. File transfer services are offered by most Internet access providers. There are a few exceptions, such as Sprint, and WestNet.

E-mail is the most commonly used application on the Internet. E-mail is supported by a wide range of delivery agents (or text editors) and delivery mechanisms on the Internet. Examples of delivery agents include Unix-based systems such as Berkeley Mail program and public mail programs such as Pine and Mush. Examples of e-mail delivery mechanisms are SMTP, a TCP/IP based-protocol, and X.400, an OSI-based e-mail program. SMTP is by far the most common e-mail delivery mechanism and is used for transferring messages between different hosts. One of the advantages of SMTP to the technical telecommuters is that SMTP protocol transactions are transparent to the technical telecommuter. While SMTP and X.400 cannot communicate directly, Internet providers, particularly the national Internet providers, provide gateways between the delivery agents of both protocols. Multipurpose Internet mail extension (MIME) is a recent extension to basic Internet mail. E-mail services are offered by most Internet access providers. Exceptions include Sprint and SURANET.

Bulletin board services are extensions of the e-mail service. This service provides true public messaging, allowing users (including the technical telecommuters) to send messages that can be accessed by other Internet members using similar bulletin board systems. The best known bulletin board service on the Internet is *USENET*. In addition to USENET, there are other bulletin boards, including gateways to commercial bulletin boards such as Compuserve.

Gopher services allow the telecommuter to browse information across the network without having to login or know in advance where to look for information. The gopher system offers information as a simple hierarchic system of menus and files. Gopher services are offered by CERFNet, CICnet, JvNCnet and NEARNet.

Directory services provide technical telecommuters with the ability to locate information about users, services, or service providers. Directory services are often divided into white pages and yellow pages services. The white pages service provide information about individual users, while the yellow pages service provides information about services and their providers. One of the earliest directory services is the WHOIS service, a basic user directory service originally conceived to track key network contacts for the early ARPA Internet.

The X.500 directory service, an ITU-T standard, is an Internet project that is currently under development. Although the pilot X.500 projects concentrated on the application of X.500 as a white pages service, many of the supporters of X.500 suggested that it could be used as a yellow pages service. X.500 directory services are currently offered by Performance Systems International, Inc., and SURAnet.

The archie service catalogues the contents of hundreds of online file archives. The archie service gathers together the location information, names, and other details of files and indexes them in a dedicated database. Users can then contact an archie server and search this database for needed files. The archie service is accessible through a range of Internet services, such as Telnet and e-mail. The prototype archie service now tracks over 2 million filenames on over 1,200 sites around the world. There are currently at least 12 archie servers around the world. Archie services are currently supported by the major Internet service providers with the exception of ANS, SURANet, and WestNet.

Access to online library catalogues is currently offered by the Internet for many libraries. To access these catalogues, the technical telecommuter needs to establish either a Telnet session or rely on a gopher service. Accessing these catalogues is, however, not an easy task due to the lack of functional yellow pages.

The Internet provides telecommuters with access to online databases such as Dialog, Dow Jones, and Lexis/Nexis. The Internet also provides telecommuters with access to the online library world, and millions of public-domain shareware and freeware programs.

7.3.5 Logical Access To the Internet

7.3.5.1 The Time-Sharing Host Solution

A telecommuter can access the Internet by obtaining an account on a time-sharing host that is located at the service provider's premises [6]. In this case, the TCP/IP protocol stack resides on the time-sharing host. To access this host, the technical telecommuter needs to have a workstation and a modem. To establish a connection with the time-sharing host, the technical telecommuter establishes a dialup connection to the terminal server, which is attached to the Internet access provider's host. The dialup connection operates at speeds ranging from 2.4 Kbps to 28.8 Kbps. (While POTS is currently the predominant access method for technical telecommuters, ISDN is emerging as an alternative method.) This dialup connection enables the technical telecommuter to dial into a terminal server that establishes a session with the time-sharing host.

This host in turn establishes a connection across the Internet with the target host. A file transfer then occurs from the target host to the time-sharing host. The file is then downloaded to the technical telecommuter's workstation through a *serial file transfer protocol* such as Kermit or Xmodem.

Examples of time-sharing hosts are Internet public access sites, which are referred to as PDIALS. Another example is Freenet. These public access sites exist predominantly in urban areas, such as New York City and San Francisco. They offer services that include e-mail, USENET, anonymous FTP services, remote login (Telnet), and remote navigators (e.g., gopher). Charges for access to these PDIALs range from $1 to $3 per hour [7]. In addition, the technical telecommuter who is using a modem incurs the cost of telephone charges.

Another example of public access sites are public access Unix systems, sometimes referred to as "nixpubs." There are more of these types of public access sites than PDIALs. These types of sites support the technical telecommuter with a narrower range of services than those offered by PDIALS. Examples of these services include USENET news groups, e-mail, and local-source archives. Charges by these public access sites range from $5 to $15 per month [7]. Another approach to accessing a time-sharing host is through 1-800 accessible services, such as PSILINK, which offers e-mail and may also offer access to UseNet Newsgroups and anonymous FTP sources.

7.3.5.2 Full TCP/IP Connection

Full TCP/IP connection is another option available to corporations to access the Internet. This connection can be implemented in one of two different ways: (1) a full TCP/IP suite may actually reside on the technical telecommuter's computer, or (2) a corporation may decide to establish an Internet node on the corporate side. In this case, the technical telecommuter accesses the Internet through the corporate Internet node.

The first full TCP/IP connection option enables the telecommuter to run the TCP/IP protocol applications on his or her PC, including e-mail, USENET, Telnet, FTP, and (possibly) gopher. In these cases, the technical telecommuter's workstation acts as a peer with other hosts on the Internet. This enables the technical telecommuter to move files directly from and to his or her hard disk instead of going through the two-step process associated with accessing a time-sharing host. This approach also enables the technical telecommuter to access remote file servers via network file systems (NFS) as apparent extensions to his or her file disk. Through this approach, the technical telecommuter can also support multiple sessions simultaneously in separate windows. The technical telecommuter can establish a connection between his or her TCP/IP-based computer and the Internet either on demand or on a full-time basis. TCP/IP services costs anywhere from $5 to $10 per hour.

As indicated earlier, another full TCP/IP interconnection option available to a corporation is to establish an Internet node on the corporate site. In this case, the technical telecommuter can access the TCP/IP-based corporate host through a modem, a dialup line, and a serial line protocol. An example of the serial line protocols is the serial line internetworking protocol (SLIP). This is a networking layer protocol (level 3 of the OSI reference model) that is widely supported by midlevel service providers and campus computing centers.

7.3.6 Internet Access Speeds

A number of access methods and speeds are available to the telecommuter, including (1) a dialup connection, (2) X.25 dialup, (3) toll free dialup, and (4) ISDN.

Dialup connections are offered by the major Internet service providers with the exception of SURANet. These connections range in speed from 9.6 Kbps to 28.8 Kbps. Toll free dialup is offered by the major Internet service providers with the exception of SURANet. These connections range in speed from 9.6 Kbps to 28.8 Kbps. ISDN access to the Internet is emerging. For example, Performance Systems International (PSI) began providing ISDN BRI in 30 cities [8]. While telecommuters can rely on these methods to access the Internet, corporate locations can access the Internet through several other methods, including frame relay and SMDS.

7.3.7 Strengths and Limitations of the Internet

The Internet represents an appropriate networking and communications solution to the technical telecommuters for several reasons, including its low cost, its extensive information resources and the wide range of applications that it supports. Each of these issues is examined below.

7.3.7.1 Low Cost

This is one of the key strengths of the Internet solution [9]. As discussed above, the Internet cost to the telecommuter may range from $1 to $10 per hour in addition to telephone charges. As a result, the Internet solution is more cost effective than the VAN solution described later in this chapter. While VAN providers' charges are based on level of activity (e.g., number of messages or number of files transferred), Internet charges are insensitive to the level of activities. The low cost of access to the Internet enables large companies as well as small companies to establish business communications links with other existing potential buyers and suppliers.

7.3.7.2 Extensive Information Resources

This represents another major source of strength of the Internet as an application solution for the technical telecommuter. Through the Internet, the technical telecommuter can access hundreds of libraries around the world, as well as library catalogues and full text-service delivery services. The Internet also enables the technical telecommuter to access a wide range of government agency information, such as space shuttle updates. In addition, the Internet provides participants with a wide range of public domain software and freeware documents, databases, images, and other files that they can rely upon in their technical development efforts.

7.3.7.3 Wide Range of Applications

Another major source of strength of the Internet is the wide range of applications that it supports, including logon services e-mail, file transfer, host-to-host communications, and directory services. These applications enable a technical telecommuter to establish a dialog with other researchers, and scientists around the globe. In addition, new tools such as *electronic white boards* enable Internet members to share in real-time notes about specific technical topics. The ultimate result of this increasing level of interactions among Internet participants (including the technical telecommuters) is the speeding of the process of scientific research and product development, and a strengthening of business ties among members of the Internet community. Members of a "virtual corporation" could also use those applications.

The Internet, however, has several limitations that technical telecommuters and IS managers need to take into account in deciding whether to join the Internet community.

First, technical telecommuters need to be aware of the fact that many parts of the Internet disallow information transmitted for commercial gain. This is an important consideration that IS managers need to take into account in selecting an Internet provider if they want to select a provider that supports unrestricted traffic.

Second, telecommuters need to know that while the Internet-based e-mail has a consistent method for assigning addresses, the method is hard to remember and to use in finding addresses.

Third, telecommuters need to consider the ease of use of Internet services. While e-mail use is a fairly straightforward process, other applications are more complex. Consequently, depending on the application, the Internet may not be the most appropriate solution for some nontechnical telecommuters.

Fourth, telecommuters need to consider that they cannot be guaranteed a certain throughput across the Internet nor a consistent reliability level. This is because the Internet, while serving thousands of organizations and millions of individuals, lacks any mechanism for reserving bandwidth. In addition, the Internet is made up of many networks. The establishment of common reliability levels requires complex interactions among a large number of Internet providers.

Fifth, and probably the most important limitation of the Internet, is its security aspects. Several Internet security violations have occurred lately. These violations, which were committed by college or high school students, involved unauthorized password use. This resulted from the emergence of programs collecting names and passwords on the network. Fortunately, several solutions are becoming increasingly available to corporations that want to secure Internet access. These solutions differ in terms of the degree of security they provide to a corporation. Two of the increasingly popular solutions are described here.

1. *Screening router firewall*: This solution involves using routers configured by using native access rules to filter packets at the IP port level. This represents a good first step for a company implementing security measures. The benefit of this solution is that it may not require additional investment on the part of the corporation since the router may be a part of the installed base. This solution, however, lacks customization and audit capabilities.
2. *Application-level firewalls* provide a higher degree of security than router firewall. The implementation of this method is accomplished by enabling users to *telnet* to application-level prompts. The benefit of this solution is its audit capability. Products that support this solution, however, can be expensive and they are relatively new. Another limitation of this solution is that it requires administrative commitment. Examples of companies providing this solution are Raptor, DEC, ANS, and Trusted Information Systems.

Neither solution is foolproof, and given the security risks of the Internet, telecommuters, whether working at home or in the office, should not use the Internet for mission-critical applications until better security measures are found.

7.4 THE VAN SOLUTION

VANs provide the technical telecommuters who need to communicate with their customers, suppliers, and corporate offices with an alternative to the Internet. VANs are X.25-based networks. Packet switching, a technology that first arose in the mid-1960s, affords statistical allocation of bandwidth. Information is exchanged as blocks of limited size or *packets*. At the source, files (and other types of data blocks) are partitioned into an appropriate number of packets; packets are transmitted across the network and are reassembled at the destination to reconstitute the original file (or data block). Multiple users can share network resources, thereby lowering the costs. Packet-switching service can be obtained via a privately owned network or via a public packet-switched carrier, including a VAN. Packet switching has been standardized according to the ITU-T (International Telecommunications Union-Telecommunications, formerly CCITT) Recommendation X.25, first published in 1976 (the recommendation was significantly revised in both 1980 and 1984; minor revisions have taken place since). Packet-switching's throughput has traditionally been limited to around 9.6 Kbps (more recently, 56 Kbps, particularly in the backbone) and, hence, is not ideally positioned to support all of today's data-intensive applications. Some users, however, could still employ it—for example, for e-mail and store-and-forward applications. Packet-based networks are typically priced on a timed-usage basis, and are therefore sensitive to data volume; they are, however, insensitive to distance and are only slightly sensitive to the number of sites added. This service is typically provided by VANs at the national level and by some LECs at the local level.

VANs provide the subscribers with a number of services, including e-mail, database access, X.500 directory services, integrated messaging, SNA support, router services, closed user groups, and other enhanced services. Each of these services [10] is described below.

7.4.1 VAN Services

7.4.1.1 Electronic Mail

Most VANs offer electronic delivery of messages on a 24-hour basis. Most VANs also support multiple address and broadcast messages in addition to electronic indexing and storage. MCI offers extensive e-mail capabilities through MCI Mail. AT&T's Easylink network also has strong e-mail capabilities through its store-and-forward application architecture. X.400 messaging services support e-mail and other messaging based on the X.400 standard. X.400 enables multiple electronic mail systems to communicate with each other. BT and Sprint are high-end international VANs that support X.400 standards extensively in international communications.

7.4.1.2 Database Access

This application enables users, including technical telecommuters, to access public and private databases maintained by the VAN. Examples of these private databases include Sprint's America Online and CompuServe information service.

7.4.1.3 X.500 Directory Services

X.500 directory services are offered by VAN providers either to serve a single network (e.g., an e-mail network) or to provide members of different networks with information about one another. The two major VAN providers that support X.500 are AT&T and GEIS.

7.4.1.4 Integrated Messaging Services

Integrated messaging services are offered by several VAN providers through various architectures. For example, AT&T's Unified Messaging Architecture (UMA) allows e-mail, fax, and EDI messages to be left in a single mailbox. Similar capabilities are supported by BT's Global Network Services and Sprint-Net. This capability enables the users to access multiple applications from a single logon and through a single access line. Other VAN providers pursue a different approach to messaging integration. For example, GEIS was developing software that provides a single PC-interface to EDI and e-mail.

7.4.1.5 SNA Support

VAN providers vary in terms of their expertise in certain architectures, such as IBM's SNA environment. For example, Advantis, which was formed in 1992 as result of the merger of the VANs of IBM, Sears Roebuck, and Cox, focuses on implementing and maintaining IBM SNA environments.

7.4.1.6 Router Services

These services enable users to link LANs over VAN providers routers and backbone networks using protocols such as the TCP/IP. Over 75% of the VAN providers can support router networks. Examples of these are Infonet and Sprint.

7.4.1.7 Closed User Groups

This service provides restricted access to the network. Members of each group can only communicate with other members of the group. This feature can be utilized by companies that want to restrict their e-mail to data communications between a select group of companies or individuals.

7.4.1.8 Local Error Protection

This VAN feature provides error detection and protection in the access portion of the network. This feature is particularly important for asynchronous protocol transmission, which does not provide native error handling procedures.

7.4.1.9 Other Enhanced Services and Features

Other enhanced services and features offered by most VAN providers include dial back-up, protocol conversion, network management, and digital multiplexing [11]. In addition, some of the larger VAN providers offer outsourcing services.

7.4.2 VAN Providers

There are currently over a dozen major VAN providers in the U.S. These vendors generate VAN revenues that were estimated to be worth more than $3.3 billion in revenues in 1992. According to Northern Business Information [12], the four major leaders of VAN services in the United States in 1992 are BT North America (22%), Advantis/IBM (17%), SprintNet (25%), and Compuserve (17%). The characteristics of these major domestic VANs, including number of domestic points of presence, number of foreign nodes, access methods, protocols supported, network management services, pricing, and broadband migration strategy are shown in Table 7.3. The three international VAN providers with the most market share are BT, GEIS, and AT&T.

7.4.3 VAN Access Alternatives

A technical telecommuter can access the network through a number of methods, including the following.

1. *Asynchronous public dialup* [13] service supports any commonly used synchronous terminal at speeds up to 19.2 Kbps. A technical telecommuter can access the public VAN network through the nearest VAN network node. Local dialup ports provide toll-free access to a network node. Technical telecommuters can receive a list of local access locations from the VAN provider. Most VANs also offer 800 numbers INWATS for outlying areas. These numbers are subject to local surcharges.
2. *X.25 dialup* service supports asynchronous and synchronous transmissions and provides end-to-end error detection and correction.
3. *Private dialup* service permits exclusive use of a dialup port on a continuous basis. Features offered by this service are similar to those offered by public dialup services (asynchronous and X.25). The difference, however, is that the telecommuter (or his or her corporation) is charged a flat monthly fee instead of an hourly connect charge.
4. *ISDN* access is now offered by some VAN providers.

The corporate office can be interconnected to the VAN through a *dedicated terminal access* method, which refers to leased lines that interconnect the corporate office with a dedicated port on the VAN. These leased lines have higher speeds and reliability than switched dialup connections. Corporations are charged a flat fee for this service.

7.4.4 Strengths and Limitations of VAN Services

VAN services offer telecommuters several benefits. First, VAN services are easy to use. The major services offered by companies such as Compuserve feature

Table 7.3
Characteristics of Major Domestic VAN Providers

Network	*Advantis/IBM*	*CompuServe*	*Sprint*
Service family	Advantis	Packet services	SprintNet
Domestic POPs	276	370	526
Foreign nodes	384	9	343
Countries reached by X.75	90	—	117
Access methods:			
Public, private dial	9.6 Kbps	9.6 Kbps	9.6 Kbps
Dedicated	1.5 Mbps	5.6 Kbps	256 Kbps
Frame relay	—	1.5 Mbps	256 Kbps
Other			Dial out, single number 800, personalized access
Protocols supported	Asynch, BSC, SNA, and TCP/IP	Asynch, SNA	Asynch, BSC, SNA, and TCP/IP
Enhanced Services:			
E-mail, X400	✓	✓	✓
DB retrieval	✓	✓	✓
Router service	✓	✓	✓
Managed network service	✓	✓	✓
Network management services	Reports, SNMP	Reports, SNMP, terminal	Reports, terminal
Pricing			
Charge per port per month	$750–$4,800	$350–$3,000	$615–$1,525
Usage fee per K-character	—	$.03–$.05	$.02–$1.40
Connection time charge/hr	$4.45–$5.50	$2.80–$5.75	$1.17–$9.00
Volume discount	✓	✓	✓
Broadband migration strategy	ATM (1995)	ATM (1994) ATM/frame relay connectivity	X.25/frame realy interconnectivity, frame relay/SMDS, SONET, ATM

Sources: Datapro, *Network World*, and Northern Business Information.

colorful, graphic interfaces and easy access to help screens. Second, VAN service providers offer additional enhanced services that the Internet does not support, such as outsourcing services and network management. These services are appropriate for corporations that are contemplating outsourcing some IS functions. Third, VAN services support several applications that the Internet does not, such as EDI and online transaction processing (OLTP).

On the other hand, VANs have a number of limitations when compared with the Internet providers: (1) the number of technically oriented databases on VANs are much more limited compared to the Internet; (2) VAN prices would be higher to most telecommuters because VAN charges are usage sensitive and technical developers spend many hours online; and (3) the number of subscribers to VANs is lower than the number of Internet subscribers. In addition, the Internet community is growing much faster. While VAN subscribers are growing at 20% to 30% per year, the number of Internet subscribers is growing at 100% per year.

7.5 VAN/INTERNET COMPLEMENTARITY

Recognizing the growing importance of the Internet, a growing number of VANs (e.g., Sprint and AT&T) are providing their subscribers with the capability to access the Internet. Figure 7.4 illustrates the Internet access through SprintNet [14].

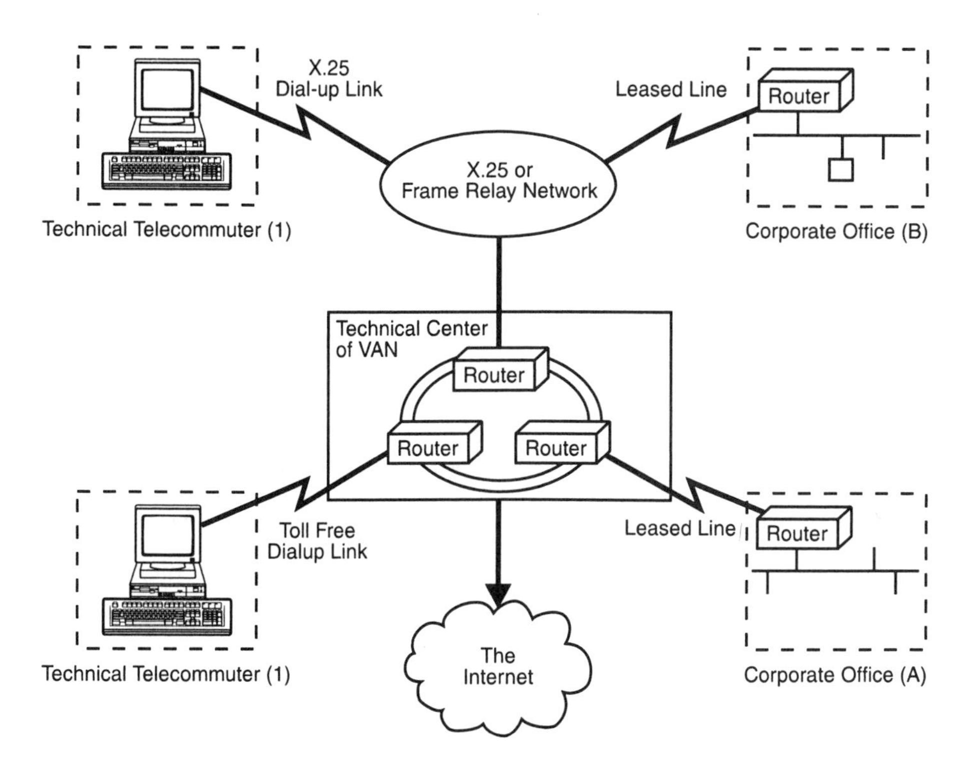

Figure 7.4 Access to the Internet through VANs. (*After*: [14].)

References

[1] Kobielus, James, " Groupware Buyer's Guide—The Time is Ripe to Pick Groupware," *Network World*, Aug. 9, 1993, p. 47.

[2] "Lotus Notes—Setting the Pace in Workgroup Computing," product literature.

[3] "Groupware: A Major Look-Forward Driver for Bandwidth Consumption," *Strategic Focus . . . on Connectivity*, April 1993, p. 27.

[4] Bosco, Paul D., and Hans-Werner Braun, "The NSFNET T1/T3 Network, Backbone of Emerging Information Infrastructure Celebrates 5 Years of Extraordinary Growth (1988–1993)," *Connexions*, p. 2.

[5] "Internet Service Providers Buyer's Guide," *Network World*, Vol. 10, July 19, 1993.

[6] "Internet Primer," Technical notes, p. 36.

[7] Dern, Daniel P., "Start Here: How and Why To Join the Internet and Get Going," *Internet World*, Sept./Oct. 1993, p. 68.

[8] "ISDN Access Provides Cheaper Internet Links," *Computerworld*, Nov. 22, 1993, p. 57.

[9] Stecklow, Steve, "Internet Becomes Road More Traveled as e-Mail Users Discover No Usage Fees," *The Wall Street Journal*, Sept. 2, 1993.

[10] Anderson, Patty, "Value-Added Networks (VANs): Overview," Datapro Communications Series: Communications Networking Services, June 1993, p. 6201–6208.

[11] Briere, Daniele, and Cristopher Finn. "Not just X.25 anymore," *Network World*, March 22, 1993, pp. 45–52.

[12] *VAN Markets: 1993 Edition*, Dossier: U.S. Telecom Service Markets, Northern Business Information, March 1993, pp. 4–5.

[13] "VAN Services and Markets," Datapro Network Services, Aug. 1992, pp. 3050–3055.

[14] Johnson, Johna Till, "Linking Corporate Users to the Internet," *Data Communications, Jan. 1993, p. 58.*

8 Regional Telecommuting Centers

This chapter describes a corporate alternative to the individual telecommuting arrangement—regional telecommuting centers (RTCs). The chapter identifies the communications requirements of RTCs, differentiates RTC communications requirements from those of individual telecommuters, and explores various networking and applications solutions. Additionally, it compares the strengths and limitations of these solutions to determine to what extent each solution meets the requirements of the RTCs.

8.1 WHAT ARE REGIONAL TELECOMMUTING CENTERS?

RTCs enable corporations to assemble a group of employees who live in the same neighborhood in a location that is in close proximity to their houses. This provides these corporations with an alternative to the individual telecommuting arrangement. There are three major types of RTCs: (1) organization-specific RTCs, (2) function-specific RTCs, and (3) geography-specific RTCs [1].

Organization-specific RTCs are facilities dedicated to a single organization or corporation. A corporation may further devote an RTC to a specific group or to multiple groups, creating a function-specific RTC. For example, a corporation may decide to devote an RTC to corporate agents. (The *hoteling* concept described in Chapter 1 fits this definition of an RTC.) Geography-specific RTCs, on the other hand, are facilities that are shared by multiple organizations or corporations. In this case, the individual organizations that are participating in the RTC share the cost of setting up and operating the RTC.

An example of a geography-specific RTC is the neighborhood telecommuting center of Riverside County, California [2]. The creation of this RTC in the State of California has been partially stimulated by clear air regulations imposing fines of as much as $25,000 per day on employers that do not cut down on traffic congestion and lengthy commutes. This RTC was opened in

November 1992 as a result of a three-way partnership among the State of California, the Riverside County Transportation Commission, and several corporations, including TRW, Southern California Edison, and Pacific Bell. The center is the largest geography-specific RTC in the country and includes 40 telecommuters. These telecommuters perform various functions, including customer service, scientific research analysis, claims analysis, and data entry. Each of the RTC participants telecommutes one to two days per week, on average. The average commute for each telecommuter is 10 minutes. Each company provides equipment for its employees and pays for their long-distance telephone charges, modem, fax, and photocopy charges. Riverside provides private offices, cubicles, furniture, and a conference room. Pacific Bell provides Centrex services, voice mail, secure data transmission, a lunch room, and an exercise room. According to a "neighborhood telecommuting workcenter" report of Riverside County, the average commute dropped from 2 to 3 hours per day to 10 minutes. Another benefit provided by the center is that people start early (around 5 AM) [2]. By arriving early, the telecommuters can take the afternoon off.

8.2 BENEFITS OF AN RTC

RTCs provide corporate employees, employers, supervisors, and labor unions with several benefits. RTCs share some of these benefits with the individual telecommuting arrangement. In addition, RTCs have their unique benefits.

8.2.1 RTC Benefits to Employees

RTCs improve employee morale by enabling them to work more flexible hours, by enabling them to spend more time with their families rather than on the road, and by eliminating the potential psychological isolation associated with work-at-home arrangement. RTCs also reduce the employees' costs associated with public transportation and/or purchasing, maintaining, and fueling a car. In addition, RTCs improve employees' physical health by reducing their exposure to air pollution.

8.2.2 RTC Benefits to Employers

Employers can gain several cost benefits by establishing RTCs. RTCs can reduce the employer's administrative costs associated with turnover, including the cost of recruitment, relocation, insurance, and reimbursement for commuting. Another major benefit of RTCs to employers is that employers do not incur the penalty fees associated with noncompliance with government air quality mandates. In addition, RTCs enable employers to gain economies of scale by allowing them to share several customer premise equipment.

In addition to cost benefits, RTCs enable employers to utilize untapped resources, including over 30 million disabled Americans who are available to join the workforce, given the appropriate working conditions. By moving the workplace closer to their homes, disabled Americans are more likely to join the workforce. RTCs also enable employers to recover more quickly from disasters by providing them with a more distributed employment arrangement, assuming that these RTCs are linked with regional offices and headquarters through a robust communications network. This is particularly important in earthquake-prone areas like California.

8.2.3 RTC Benefits to Supervisors

RTCs are beneficial to supervisors by providing them with the ability to monitor their employees more closely through periodic visits to the RTCs. In particular, these visits address the concerns of traditional supervisors who measure the performance of their subordinates according to traditional standards (e.g., number of hours spent in the office, absenteeism, etc.).

8.2.4 RTC Benefits to Unions

RTCs also reduce the concerns of the unions about the working conditions of their members by enabling union leaders to visit these RTCs [3].

8.3 COMMUNICATIONS REQUIREMENTS OF RTCs

Communications requirements of RTCs are driven by the communications requirements of two major groups: the individual participants in an RTC and the IS managers.

The communications requirements of the individual participants in an RTC depend upon the functions that they perform. They may be marketers, technical developers, administrators, or agents. The requirements of these different groups of telecommuters are similar to those requirements that have been described in Chapters 3 through 7, although the economies of scale enable other solutions to be employed.

While each participant in an RTC is interested in his or her individual communications requirements, IS managers need networking solutions that address the *aggregate* communications requirements of these participants. In selecting RTC networking solutions, IS managers take into account a number of factors, which are more extensive than those factors that they take into account in providing network solutions to telecommuters working at their individual homes. These factors include the geographic extent of network connections, the RTC application mix, the degree of vendor support for the network solution being considered, the service cost, the quality of service (QoS), and the standards. Each of these factors is examined below.

8.3.1 Geographic Extent of RTC Network Interconnections

IS managers need networking solutions that enable RTCs to establish connections with corporate offices, suppliers, customers, or agents. These connections may be few and predetermined, or they may be fairly large and constantly changing. These connections may also be intraLATA, interLATA, or international. One measure that an IS manager may use in determining the degree of availability of a service is the number of points of presence (POPs) that support this service. The number of POPs supported by each provider has a direct bearing on the cost of dedicated access facilities.

8.3.2 The Mix of RTC Applications

Four types of applications may exist in an RTC: (1) LAN-LAN interconnection applications, (2) host-to-remote bulk data applications, (3) voice applications, and (4) real-time (as opposed to store-and-forward) video applications.

These applications differ in terms of their bandwidth requirements and their degree of burstiness. LAN-LAN interconnection applications are, in some cases, time-variable (bursty) data transmissions and have high bandwidth requirements. Host-to-remote bulk data are time-sensitive data applications with medium to high bandwidth requirements. Real-time video is also a time-sensitive application with extensive bandwidth requirements. Voice is another time-sensitive application, but with much more limited bandwidth requirements. Depending upon the mix of applications, 1.5 Mbps may be an adequate bandwidth to support the aggregate communications requirements of the telecommuters.

8.3.3 Vendor Support

IS managers prefer the selection of services that are widely supported by equipment vendors, including local exchange carriers, and interexchange carriers. In addition, IS managers are increasingly considering the option of an alternate access provider.

8.3.4 Service Cost

IS managers place a high value on RTC networking solutions that minimize the recurring costs (e.g., number of leased lines, the cost of access lines) and nonrecurring costs (e.g., installation charges). Other cost considerations include network management cost, including the cost of designing, maintaining, and upgrading a service. To minimize cost, IS managers typically value *integrated access solutions* that can consolidate multiple types of traffic (voice, bulk data, bursty data, and video) on a single access line. To further minimize cost, IS managers also place a high priority on solutions that can retain their existing infrastructure, which includes wiring, hardware, and software. For example, if an IS manager selected and implemented a frame relay network, he or she would prefer to select a telecommuting networking solution that is compatible with the frame relay network.

8.3.5 Quality of Service

This encompasses several factors, including the ability of a service to guarantee the delivery of the nominal bandwidth specified by the service provider, the range of bandwidth supported by the service, response time, delay, and efficiency (packet loss) [3]. The quality of service (QoS) may also depend upon how secure the service is. Some IS managers may require stringent security solutions that are network-based (instead of corporate-based).

8.3.6 Standards

IS managers prefer networking solutions that are supported by either formal or de facto standards. This increases the set of potential sources for the equipment, thereby increasing competition and lowering costs.

8.4 RTC NETWORKING SOLUTIONS

Currently, there are several networking solutions that meet at least some of the IS manager's requirements mentioned above. Seven of these solutions are explored next. (These solutions, operating at DS1 speeds, provide higher speed to RTCs and other corporate offices than those offered to the telecommuters who work out of their homes.) These solutions include the following.

1. Private T1 lines;
2. Fractional T1 services;
3. Switched/dialup services;
4. SMDS;
5. Frame relay;
6. ATM;
7. Integrated T1 access.

For each solution, key technical characteristics are described, RTC equipment requirements are enumerated, key applications are identified, network service providers are identified, and key strength and limitations are highlighted.

8.4.1 Private T1 Lines

A private T1 line is a dedicated service, which provides a communication link between two locations through the establishment of a physical connection supporting a data rate of 1.544 Mbps. Private T1 line services are currently widely available in the United States and are offered by LECs, AAPs, and IXCs. Two versions of private T1 lines are currently offered by network service providers: nonchannelized private networks and channelized private networks.

8.4.1.1 Channelized Private T1 Lines

Time division multiplexers (TDM) are currently the most prevalent types of equipment supporting the channelization of private T1 lines. (Figure 8.1). A TDM can combine voice, data, and video signals and then send those signals sequentially at fixed intervals. This is accomplished by subdividing a DS1 channel into 24 DS0 channels (64 Kbps). Multiple channels can be allocated to one type of traffic (e.g., voice). Sophisticated multiplexers can channelize the aggregate link into subchannels of any speed, not only 64 Kbps.

8.4.1.2 Nonchannelized Private T1 Lines

These private lines carry an aggregate 1.544 Mbps in an unslotted manner. In this case, the full T1 payload is used for the transmission of traffic, which could be data or video. To deploy an unchannelized T1 solution in support of data applications, a router is used at the RTC location as well at the corporate location. If the key RTC application is video, a video codec replaces the router (Figure 8.2).

The *supply structure* of T1 private lines has changed significantly since the days of AT&T's divestiture. While the network service providers in the early 1980s were limited to two groups, the local exchange carriers and the interexchange carriers, currently an increasing number of alternate access

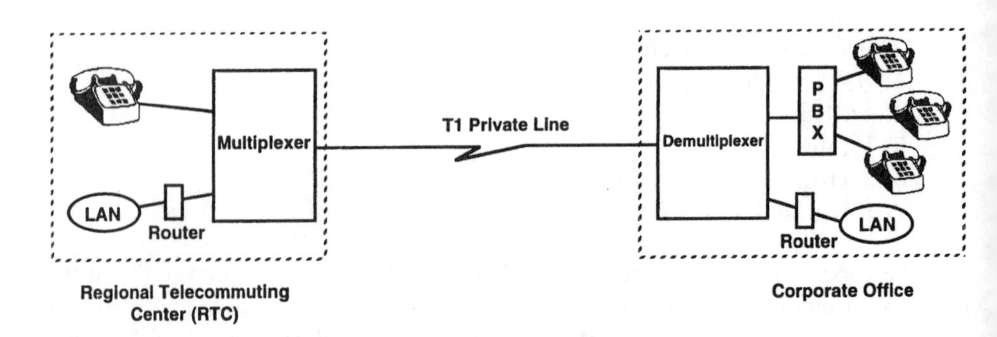

Figure 8.1 The RTC/corporate office data communications link DS1 private line solution.

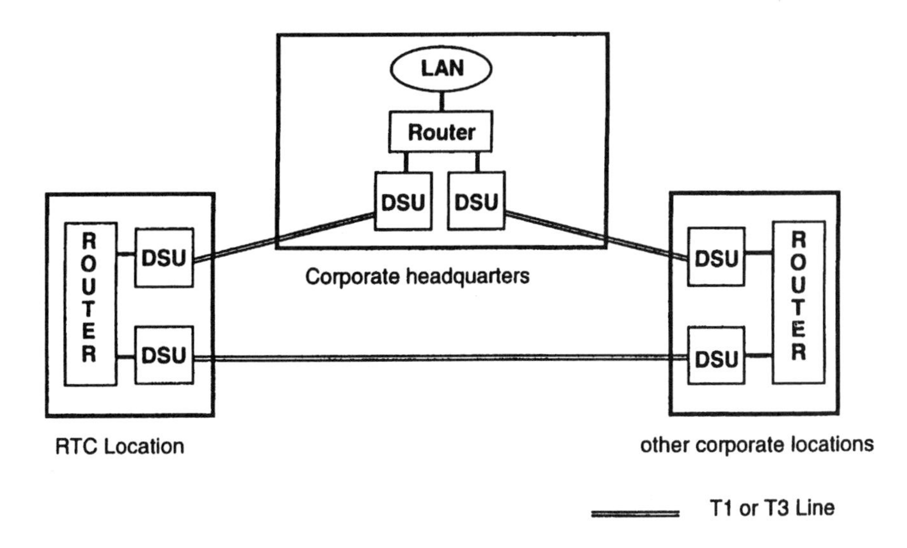

Figure 8.2 The RTC/corporate office data communications link unchannelized private line solution.

provider offer private T1 lines. Two well-known AAPs are Teleport and Metropolitan Fiber Systems (MFS); they currently control over 60% of the AAP market. As a result of the increasing competition, T1 private line prices have been dropping.

The private T1 solution provides the RTCs with several benefits, including security, adequate bandwidth for basic business applications, an established supply structure, reliability and scalability. Each of these factors is examined next.

1. *Security*: Private T1 lines are perceived by IS managers to be secure since they are dedicated to individual customers. However, tapping into copper wire is not very difficult. Fortunately, an increasing number of these lines are carried over fiber facilities.
2. *Adequate bandwidth*: T1 private lines provide RTCs with adequate bandwidth for basic traditional applications. This bandwidth can be subdivided into multiple channels. Each channel or group of channels can support the various type of traffic, including voice, signaling and features, data, and video.
3. *An established supply structure*: Private T1 line services are established: telephone operations and transport standards exist; it is widely supported by the LECs, the IXCs, and an increasing number of AAPs. In addition, the increasing competition is driving down the cost and making the service even more appealing.

4. *Reliability*: T1 private lines are reliable and proven. Information transport in dedicated bandwidth partitions guarantees throughput with no lost packets and processor bottlenecks.
5. *Scalability*: T1 private lines are scaleable either by adding other T1 lines or by moving up to T3 systems, allowing for flexible network configuration.

Private T1 solutions, however,are not optimal in all situations. One of the drawbacks of this solution is that T1 economics become less appealing to IS managers as the number of corporate locations requiring interconnection increases and as the corporate need for the establishment of a mesh topology increases. TDM-based private line solutions are also bandwidth inefficient because they allocate bandwidth to devices that have nothing to send. Another shortcoming of T1 private lines is that the bandwidth of T1 may be inadequate to support the requirements of the technical telecommuters, particularly for imaging and CAD/CAM applications.

8.4.2 Fractional T1 Service

Fractional T1 (FT1) services are becoming increasingly available by telephone companies, particularly IXCs, which offer national inter-LATA services. FT1 services provide users with an economic alternative to private T1 lines. Fractional T1 services provide RTCs with bandwidth in n × 64 Kbps increments. RTCs participants can have access to IXC fractional T1 services through an LEC T1 line.

The FT1 solution provides RTCs with cost savings in comparison with the full T1 solution by offering fractions of T1 bandwidth. It is priced at a fraction of the cost of a T1 line. For example, an FT128, supporting 128 Kbps, is generally about one-twelfth of the cost of a full IXC T1 line. FT1 service is also appropriate for RTCs that have requirements for nationwide networks since it is supported by the three major interexchange carriers. As is the case for a T1 line, several lower speed lines can be aggregated by an RTC-resident fractional T1 mux into a single fractional T1 signal.

FT1 services, however, are not optimal in all situations for at least one of the following several reasons: (1) FT1 may not be available in all locations as an intra-LATA service; (2) RTC participants, particularly technical telecommuters, may need the full T1 bandwidth.

8.4.3 Switched/Dialup Services

Switched/dialup services are circuit-switched-based services that support speeds ranging from 56 Kbps to T3 and carry voice, video, and data traffic. While with private lines RTC participants need multiple lines to establish connections with multiple locations, with switched/dialup service, RTC participants only need one access line to communicate with multiple locations (Figure 8.3). To establish a connection with other locations, an RTC participant dials the number that he or she needs to reach (the addressing is often based on the North American dialing plan). When the circuit switch receives the call, a circuit is established for the duration of the call. This circuit is taken down as soon as the call is completed. Switched/dialup services offer RTC participants several advantages over private lines:

1. *Cost Reduction*: Switched/dialup services could be more economical than dedicated T1 lines, particularly when there are many sites to be interconnected and the session is short, because with switched/dialup services each corporate site requires only one connection to reach all other destinations. Switched/dialup services can also reduce corporate administrative costs if network connections of an RTC cannot be predicted in advance or if these connections are constantly changing. Switched/dialup services also reduce corporate administrative costs by putting the problems of leasing most of the circuits in the hands of the carrier [4]. Switched/dialup service also simplifies network administration because it uses the E.164 ISDN numbering plan. Switched-dialup services may also be more economical than private lines because they are usage-sensitive while private line services are not. The economics depend on usage levels. Usually for usage of about 30 minutes or less per day, the switched service is cheaper; beyond 30 minutes per day, the dedicated service is generally cheaper.
2. *Support for multiple applications*: Switched/dialup service is well-suited for RTCs that have extensive needs for voice communications, video transmission, and certain data communications applications, such as bulk data transfer.
3. *Extensive bandwidth capabilities*: Switched/dialup services provide the IS manager with bandwidth capabilities that range from 56 Kbps to 45 Mbps.
4. *Flexibility*: Switched/dialup services are flexible because they enable IS managers to include RTCs in their corporate networks without having to justify the expense of an end-to-end dedicated line.
5. *Migration capabilities*: Switched/dialup services provide RTC participants with a migration path to ISDN.

While switched/dialup services can provide an RTC with several benefits, they do have several limitations: (1) they are not widely available yet for speeds above 56 Kbps (this situation may change in the next few years, as RBOCs are increasingly deploying circuit switched services at T1 speeds); (2) as a circuit-switched technology, switched/dialup services have long call setup times (3 seconds); and (3) switched/dialup services cannot properly support all LAN interconnection applications.

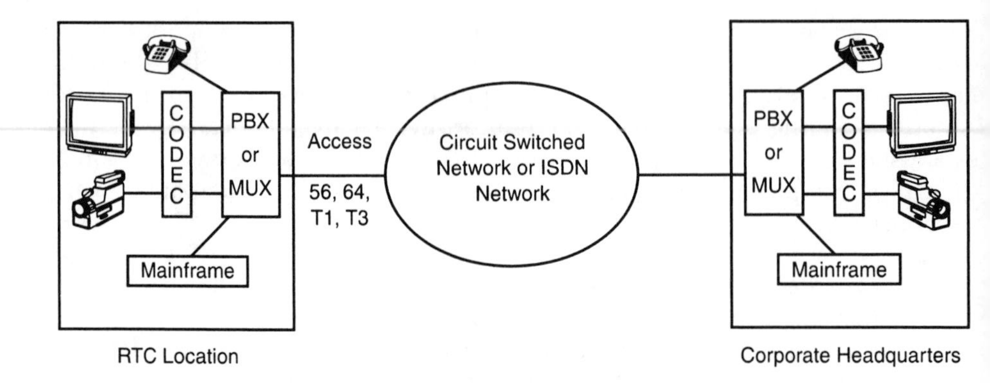

Figure 8.3 The RTC/corporate office data communications link switched/dialup solution.

8.4.4 SMDS

SMDS is a high-performance, public, connectionless (datagram transfer mode) service that was developed by Bellcore (Bell Communications Research) and is based on a subset of the American IEEE 802.6 metropolitan area network (MAN) standards.

Logical access to SMDS is established through a subscriber network interface (SNI), which may provide either a DS3 access path (with multiple access classes) or DS1 access path. For a DS3 access path, single or multiple-CPE access arrangements are available. SMDS interface protocol (SIP) is based on the IEEE 802.6 standard distributed queue dual bus (DQDB) connectionless protocol [5]. This protocol operates at the MAC level of the data link layer. As a result, SMDS can be supported by various higher layer protocols, such as TCP/IP and DECNET (Figure 8.4). SMDS may also be available in the future via ATM access.

SMDS supports five access classes. One of these access speeds can be supported by DS1 access line, while the other speeds (4 Mbps, 10 Mbps, 16 Mbps, 25 Mbps, and 34 Mbps) can be obtained by using a portion of a DS3 line. SMDS is now available at 64 Kbps.

The SMDS addressing scheme complies with the ITU-T E.164 plan for ISDN/BISDN "telephone numbers," which enables SMDS to provide end users with smooth transmission to ATM. SMDS has several addressing features, including multiple addresses per interface and authentication of address, in each packet. In addition, an SMDS customer can choose a list of addresses that he or she wants to communicate with; all other addresses are blocked.

In addition to its addressing capabilities, SMDS enables RTCs to establish *logical private networks,* which could be used by RTCs to establish connectivity with a select number of corporate locations, suppliers, and/or customers.

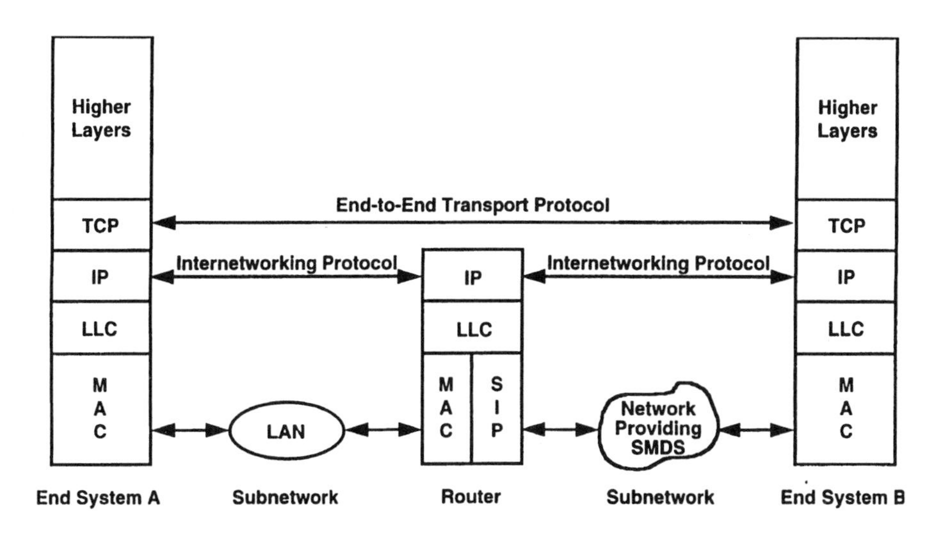

Figure 8.4 SMDS.

SMDS also has multicasting capabilities, which are similar to the broadcast/multicast capabilities of LANs. These capabilities are useful for address resolution, router updates, and resource discovery.

SMDS also has several security features, including *source address validation and address screening*, which block communication between RTCs and unauthorized SMDS interfaces. In addition, SMDS offers RTCs *low delay* (5-10 msec for a DS3-DS3 link), *high throughput* (95% of packets are delivered in less than 20 msec), *low error rates, high reliability, and availability* (99.9 %).

RTCs can deploy SMDS either as a new service or as a replacement of private lines. To deploy SMDS, an IS manager of an RTC needs to order the service from the local exchange carrier, install an FT1/T1/T3 SMDS access line from the RTC to the service office of the local exchange carrier, purchase a router or an SMDS upgrade for an existing router, and install a DSU/CSU and performance monitoring equipment.

SMDS can be selected in support of one or more of the following RTC applications.

1. *LAN-to-LAN interconnection*: SMDS is effective in interconnecting an RTC-based LAN with other LANs located in other customer locations as well as in locations of customers, suppliers, and customers.

2. *Channel extension*: SMDS can interconnect a mainframe located in a corporate office with peripherals located in an RTC, at the channel level. Through SMDS, bulk data can be transmitted from the host to peripherals.
3. *Video transmission*: While SMDS is not appropriate for real-time video transmissions, several successful trials indicate that SMDS can transmit store-and-forward video traffic.

SMDS can be purchased from the regional Bell operating companies (RBOCs), which at the time of writing are in the process of deploying the service. MCI has also announced the availability of nationwide SMDS and has published an interLATA SMDS tariff. Other interexchange carriers (e.g., Sprint) indicated that they will provide their customers with interfaces between their own networks (e.g., frame relay) and SMDS. This service is also supported by multiple router vendors (e.g., Cisco, Wellfleet), vendors of DSU/CSUs (e.g., ADC Kentrox), and customer network management (CNM) providers (e.g., Cabletron, Hewlett Packard).

SMDS provides RTCs with several benefits, including cost reduction, support for high-bandwidth bursty data applications, flexibility, and security. Each of these benefits is described below.

1. *Cost reduction*: Under certain circumstances, SMDS could be more economical than dedicated T1 lines, particularly when there are many sites to be interconnected, because with SMDS each corporate site requires only one connection to reach all other destinations. SMDS can also reduce corporate administrative costs if network connections of an RTC cannot be predicted in advance or if these connections are constantly changing. SMDS also facilitates network administration because it uses the E.164 ISDN numbering plan, which simplifies addressing to other SMDS subscribers. SMDS may be more economical for certain small organizations that have high data communications needs but cannot justify the expense of building a private network. In the long term, SMDS could also be seen as an economic service because it provides a graceful migration path to ATM technology. This comes about since SMDS and ATM have several technical similarities, including the ability to transfer data in fixed, 53-octet cells.
2. *Support for bursty data applications*: SMDS is well-suited for RTCs that have extensive needs for bursty data traffic communications at relatively high bandwidth levels. For example, if the RTC is occupied predominantly by technical telecommuters, SMDS makes sense.
3. *Extensive bandwidth capabilities*: SMDS provides the IS manager with a choice of various access classes, enabling him or her to start at a lower bandwidth level and then migrate the network to a higher access class as the communications needs of the RTC warrant the upgrade.

4. *Flexibility*: SMDS is flexible because it enables IS managers to include RTCs in their corporate networks that do not justify the expense of a dedicated line.
5. *Security*: SMDS has several appealing security features, enabling IS managers to establish virtual private networks, including an RTC. As indicated above, addresses can be screened so that only authorized destinations can receive data and only authorized sources can send data. Through the group addressing capabilities of SMDS, RTCs can broadcast data to multiple locations at once.

While SMDS can provide an RTC with several benefits as explained above, SMDS has a number of shortcomings: (1) for a few locations, SMDS is a more expensive solution than private lines; (2) IXC availability is limited; (3) LEC availability is limited; (4) it may be eclipsed in the future by ATM, which now has the support of over 600 companies, which are members of The ATM Forum; and (5) it is defined today to support connectionless packet data traffic; it cannot support POTS voice, signaling, and features (e.g., CLASS), or real-time video applications. Consequently, if RTCs are populated predominantly by marketing telecommuters who, as indicated earlier have emerging video communications applications, SMDS is not an appropriate solution unless it is complemented by other solutions that support video and voice. (*Complementarity among various solutions is explored as part of the integrated access services later in this chapter.*)

8.4.5 Frame Relay Solutions

Frame relay networks provide another data communications alternative to RTCs. Frame relay service is a data-only connection-oriented frame transport service that supports the exchange of *variable length* information between two customers' premises over assigned virtual connections. These virtual connections, which are established through statistical multiplexing techniques, can be either permanent virtual connection (PVC) or switched virtual connection (SVC). Frame relay services available on the market today are PVC-based. A PVC connection is usually established when the service is provisioned at subscription time, eliminating the need for user-to-network signaling.

Frame relay service supports access speeds of 56 Kbps, N × 64 Kbps, and 1.544 Mbps. Some vendors were considering extending the frame relay service to the DS3 level.

A frame relay network is based on multiple ANSI standards, including T1.606, which specifies user-to-network interface requirements and internetworking requirements, including internetworking with X.25 and internetworking between frame relaying services; ANSI T1.617 ANNEX D which specifies critical network management functions; ANSI T1.618-1991 (LAP-F Core) which specifies the lower sublayer of the data link layer of frame relay and is based on the core subset of T1.602 (LAPD).

The frame relay data transfer protocol defined in T1.618/LAP-F Core is intended to support multiple simultaneous end-user protocols within a single physical channel. Above layer 2, this protocol is transparent. As a result, most existing protocols (e.g., TCP/IP, X.25) can ride over frame relay transparently to the end devices [6].

Frame relay networks can support the following types of applications.

1. *LAN-to-LAN interconnection*: Frame relay can support the interconnection of Ethernet/IEEE 802.3 and token-ring type LANs.
2. *Host-to-remote bulk data*: Through frame relay assembler/dissassemblers (FRADs), frame relay can support host-to-remote bulk data transfer and other IBM-type SNA applications.

Frame relay can be implemented either as a private or public solution. To implement frame relay as a private networking solution, an RTC can either be deployed by (1) purchasing routers (or frame relay software additions to existing routers) and establishing point-to-point connections among these routers; (2) by using a customer-owned frame relay nodal processor (or a software upgrade to an existing packet switches or multiplexers), which are basically frame relay switches; or (3) as a node on a T1 multiplexer (Figure 8.5).

Participants in an RTC can access a public frame relay network by (1) installing a software upgrade to a router or a bridge, by purchasing a frame relay access device (FRAD), or by direct firmware support on a host; (2) by connecting an access line between the carrier and the RTCs' CSU/DSU; (3) by ordering the frame relay interface from a telephone carrier offering the service; and (4)

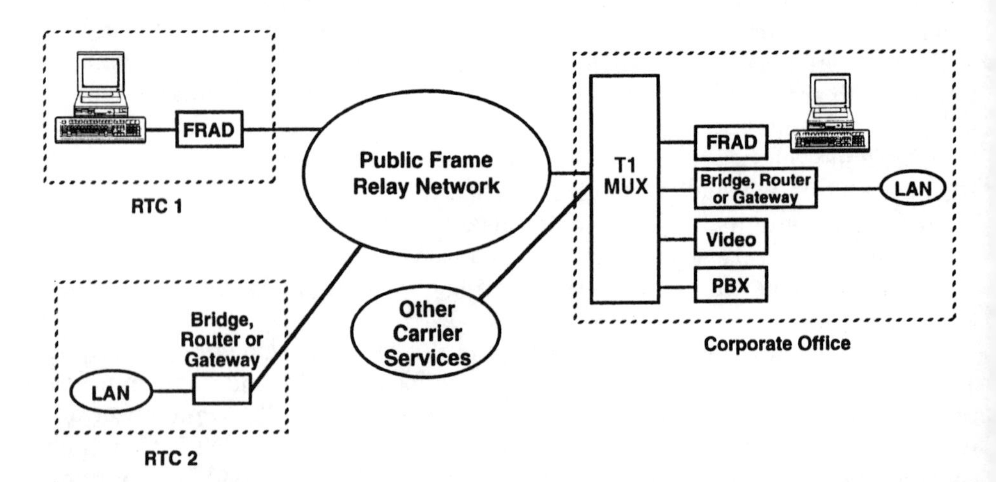

Figure 8.5 RTC/corporate office links private frame relay solutions.

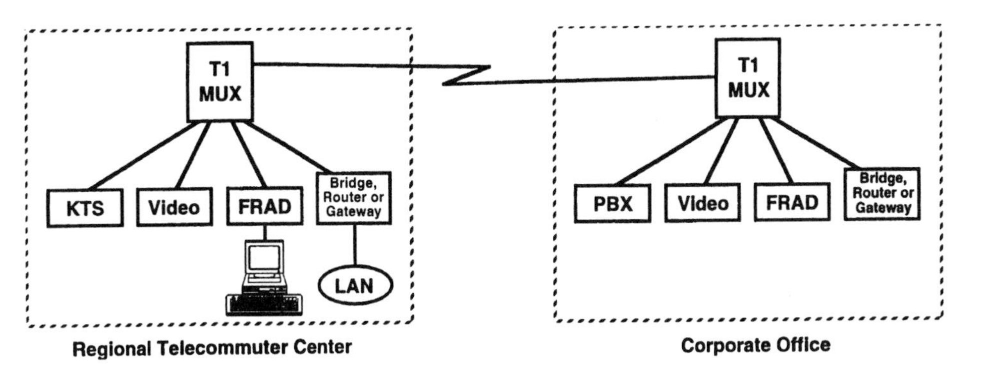

Figure 8.6 RTC/corporate office links the public frame relay solutions.

by configuring a PVC for RTC and other sites to be linked to the network (Figure 8.6). A wide range of carriers currently support frame relay, including RBOCs, the major IXCs, and several VAN providers [7].

Frame relay provides RTCs with several benefits, including cost reduction, support for bursty data applications, flexibility, and availability. Each of these benefits is described below.

1. *Economic benefits*: Frame relay offers higher transmission speeds, simplified network design, and lower operating costs than a T1 private network, SMDS, or an X.25 network. It is more economical than TDM-based T1 lines because it is based on statistical multiplexing techniques. In the long term, frame relay service could also be viewed as an economic service because it provides a migration path to ATM technology.
2. *Support for bursty data traffic*: Frame relay is well-suited for RTCs that have extensive needs for bursty data traffic communications at medium bandwidth levels. Frame relay provides the IS manager with a choice of various speeds ranging from 56 Kbps to T1 lines, enabling him or her to start at a lower bandwidth level and then migrate the network to a higher speed as the communications needs of the RTC warrant the upgrade.
3. *Availability*: Frame relay is widely available and is currently supported by IXCs, LECs, RBOCs, and VANs.

While frame relay can provide an RTC with the benefits explained above, it does have several shortcomings, including lack of support for isochronous traffic, complex administration, bandwidth limitations, inconsistent product offerings, transmission delays, poor congestion control, and packet loss.

1. *Lack of support for isochronous traffic*: Frame relay only supports packet data traffic. It cannot support voice and real-time video applications. Consequently, if RTCs are populated predominantly by marketing telecommuters, who, as indicated earlier, have video communications applications, frame relay service cannot provide a total solution to the RTC. Consequently, frame relay service is not appropriate unless it is complemented by other solutions.
2. *Complex administration*: Frame relay requires careful administration for large networks because the network administrator must define the connections for everyone on the network when users are added or changed. Network administration is becoming an increasingly important issue to large corporations as they witness two conflicting trends: network consolidation used as a cost-cutting measure and network expansion as a result the ever-continuing wave of mergers and acquisitions. These two conflicting trends further complicate network administration.
3. *Bandwidth limitations*: PVC frame relay service supports speeds of up to T1 (or E1 speeds in Europe). As result, frame relay may not be able to support the bandwidth requirements of a large number of technical telecommuters based at an RTC.
4. *Inconsistent service offerings*: Frame relay service offerings differ significantly from one vendor to another in terms of prices, range of speeds supported, network management support, support of CO-based FRAD, and pricing. These differences complicate the process of selecting and purchasing an frame relay service.
5. *Transmission delays*: frame relay service, which transmit information in small segments, have higher delay compared to private lines of the same bandwidth because of the need to store and process packet headers and trailers. Frame relay services are also prone to queuing delays because multiple sources are competing for a given trunk circuit. One source must often wait until another source is done using the trunk bandwidth.
6. *Packet loss*: Statistically multiplexed services with poor congestion control, such as frame relay service, are prone to lose packets when a network becomes highly loaded. While higher layer protocols, such as TCP/IP can detect lost packets and retransmit them, this retransmission results in decreasing throughput.

8.4.6 Asynchronous Transfer Mode

As indicated in Chapter 1, ATM[1] refers to a high-bandwidth (45 Mbps to 622 Mbps now, and more later), low-delay switching, and multiplexing technology that is now becoming available for both public and private networks [8].

1. For a more detailed description of ATM, refer to Minoli, D., and G. Dobrowski, *Principles of Signaling for Cell Relay and Frame Relay*, Norwood, MA: Artech House, 1994.

While ATM in the strict sense is simply a (data link layer) protocol, the more-encompassing ATM principles and ATM-based platforms form the foundation for a variety of high-speed digital communication services aimed at corporate users for high-speed data, LAN interconnection, imaging, and multimedia applications. ATM supports both switched (SVC) and nonswitched (PVC) connections. ATM supports services requiring both circuit-mode and packet-mode information transfer capabilities. ATM can be used to support both existing connection-oriented and connectionless services (e.g., SMDS).

ATM also supports emerging services, such as cell relay service. This is one of the key services enabled by ATM. Cell relay service can be utilized for enterprise networks that use completely private communication facilities, completely public communication facilities, or facilities that are hybrid. The first version of cell relay services available to users will be PVC-based. This service provides an economic alternative to private line services by offering users a high-performance service that can support voice (in the future), data, video, and multimedia applications at DS1, DS3 and SONET access rates. As in the case of frame relay service-PVC, RTC participants can establish connections with other corporate locations through logical connections. As Figure 8.7 shows, RTCs will be able to access an ATM-based cell relay service by relying on an ATM switch or router and a single access line. As Figure 8.8 shows, the access line can support multiple PVCs.

ATM/cell relay is an emerging service. When widely available in the next few years, ATM/cell relay can provide RTCs with several benefits, including cost reduction; support for bursty data applications, voice, video, and multimedia applications; and flexibility. Each of these benefits is described below.

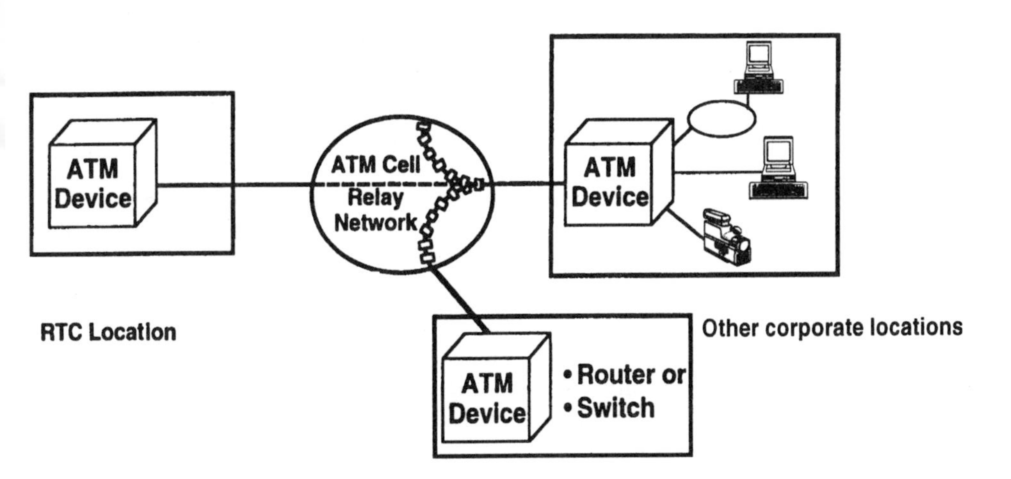

Figure 8.7 ATM cell relay service.

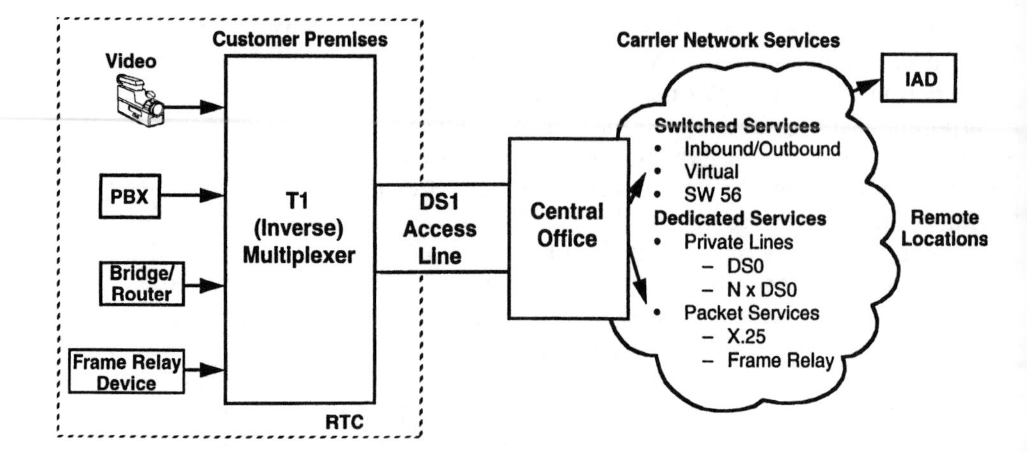

Figure 8.8 Integrated T1 access configuration.

1. *Economic benefits*: Cell relay offers higher transmission speeds than SMDS and frame relay. It is in theory also more economical than TDM-based T1 lines because a single access line is required to establish communications links between an RTC with multiple corporate locations, while with private lines, multiple access lines are required. However, the SONET, ATM, and ATM adaptation layer overhead is high, and ATM tariffs are not yet widely available.
2. *Support for data, voice, video, and multimedia traffic*: Cell relay provides RTCs with the versatility and flexibility of fixed length cells, making the service well-suited for RTCs that have extensive needs for data, voice, video, and multimedia communications. As a high-performance service, cell relay can support demanding applications, such as traffic aggregation, scientific visualization, desktop interactive videoconferencing, and distance learning. Cell relay also provides the IS manager with a choice of various high speeds ranging from T1 or SONET, enabling him or her to start cell relay implementation at a lower bandwidth level and then migrate the network to a higher speed as the communications needs of the RTC warrant the upgrade.

While cell relay can provide an RTC with several benefits as explained above, cell relay PVC has a number of shortcoming: (1) the service is the newest one to enter the market, and its availability is still low; (2) ATM service and user equipment is expected to be expensive for a number of years—just the hub/communication server connection is now $2,000 per user; (3) very little network management software has emerged; (4) originally, ATM was developed mainly to support data—video support over ATM is still being worked out;

(5) much of the recent emphasis has been premises-related issues (e.g., LAN emulation), rather than on network services; and (6) as a PVC service, cell relay requires careful administration for large networks because the network administrator must define the connections for everyone on the network when users are added or changed. SVC services should become available soon.

A variety of vendors have been introducing ATM-based products. A number of carriers either already provide services or are poised to do so in the immediate future.

8.4.7 Integrated Access

Integrated access is an emerging service that complements frame relay, SMDS, and private lines, as well as other voice services, by enabling the RTC to gain access to these services and to other dedicated services through a single access line. As Figure 8.8 shows, the integrated access line aggregates various type of customer traffic (e.g., PBX, video) and the aggregated traffic is transmitted from the customer premises to a telephone company central office (CO) where it is deaggregated and routed to the appropriate WAN services (e.g., frame relay service, PSTN, and X.25).

RTCs can gain several benefits by purchasing this service, the most important of which is cost savings. RTCs can reduce their cost of *leasing* access lines by replacing multiple access lines with a single higher speed universal access line. According to industry estimates, access charges may account for a substantial portion of the total cost of an interexchange circuit [9]. Integrated access service also reduces the cost of installing additional access lines because in many cases spare bandwidth would be available on the T1 line. In addition, integrated access service reduces installation and maintenance costs for user equipment because customers would only need to terminate and troubleshoot one high-speed access line instead of multiple low-speed lines, including some analog lines.

In addition to cost benefits, RTCs gain other benefits from implementing an integrated access service, including the following:

- The flexibility to configure a mixture of carrier services;
- Faster process of service deployment;
- Minimum interruption to existing network configurations;
- A more rapid response to changes in tariffs and service costs;
- A greater ability to handle bandwidth overflow and backup.

Integrated access service is a networking solution that could be implemented on multiple technological platforms, including ATM, SONET, TDM, channel banks, and ISDN. The three largest IXCs offer at least one technological version of this service. AT&T offers an ISDN-based version named Static Integrated Network Access (SINA), which enables the customer to bundle multiple AT&T services, including inbound/outbound services, SDN, Accunet Private lines, switched 56, and Interspan Frame Relay. AT&T also offers maintenance and management of customer-premise equipment [10].

MCI's version of integrated access supports any combination of the following services: inbound/outbound, Vnet, switched 56, private lines, and HyperStream frame relay.

Sprint's ISDN-based integrated access service supports inbound/outbound, VPN, Clearline, private lines, SprintNet X.25 service, and frame relay. Sprint also offers end-user installation and management options.

In addition to the three IXCs, the LECs offer this service as a special custom solution.

8.5 CONCLUSIONS

RTCs provide corporations with an attractive alternative to individual telecommuting. This chapter explored the benefits of RTCs to employers, employees, supervisors, and unions.

The communications requirements of RTCs are more extensive than those associated with individual telecommuters. These requirements could be met to various degrees by a number of DS1-based solutions. Six of these solutions have been examined in this chapter: private T1 lines, fractional T1 services, switched/dialup services, SMDS, frame relay, and ATM/cell relay. Table 8.1 compares the strengths and limitations of each solution.

Table 8.1
Comparison of RTC Networking Solutions

	Typical Network Connection			*Application Type*				
	Few nodes, predictable destinations	*Many nodes, predicatable destinations*	*Unpredictable destinations*	*Data*	*Voice*	*Video*	*Nonreal-time Imaging*	*Multi-media*
Private T1 and FT1 lines	•			•	•	•	•	•
Switched/dialup services		•	•	•	•	•		
Frame relay PVC		•		•			•	

Table 8.1 (continued)
Comparison of RTC Networking Solutions

	Typical Network Connection			*Application Type*				
	Few nodes, predictable destinations	*Many nodes, predicatable destinations*	*Unpredictable destinations*	*Data*	*Voice*	*Video*	*Nonreal-time Imaging*	*Multi-media*
SMDS		•	•	•			•	
ATM cell relay PVC		•		•	•	•	•	•

In addition to these services, an RTC may want to consider a complementary solution-integrated access, which enables RTCs to access some or all of these services by relying on a single access line. This solution, as this chapter demonstrated, provides the RTCs with several cost benefits.

Because of the complexity involved with selecting these high-end (and frequently expensive) services, prospective implementers may find it cost-effective to hire consultants who have installed dozens of similar systems, in order to identify and employ the best approach to avoid unnecessary mistakes. The decision to hire a consultant should take into account service availability, applications performance requirements, and costs.

References

[1] Hecht, H., "Planning for the Regional Telecommuting Centers," Enterprise Network Strategies Research Note, Strategic Planning, SPA-101-590, Gartner Group Research, Advisory & Strategic Planning Service, March 17, 1992.

[2] "Not Getting There is Half the Fun," *Telecommuting*, p. 4A.

[3] Lippis, Nicholas J., "LAN/WAN Interconnect in the Era of Broadband Internetworking," *Interop '93*, pp. 79–81.

[4] Lippis, Nick, and James Herman, "Widening Your Internet Horizons: Wide-Area Options for Internets," *Connexions*, special issue: Interop '91 Fall Companion, Vol. 5, No. 10, p. 7.

[5] "The WAN Manager's Guide to Packet Services," ADC Kentrox, pp. 20–23.

[6] "The WAN Manager's Guide to Packet Services," ADC Kentrox, pp. 17–19.

[7] Emser, Marlis, "A T1 Mux Vendor's View of Frame Relay," *Frame Relay Forum News*, Aug. 1993, pp. 9–10.

[8] Minoli, D., and M. Vitella, *Cell Relay Service and ATM in Corporate Environments*, New York, NY: McGraw-Hill, 1994.

[9] Cochran, Rosemary, "Defining the Integrated T1 Access market," *Business Communications Review*, May 1993, pp. 35–39.

[10] Gareiss, Robin, "AT&T to Offer Integrated Access," *Communications Week*, Sept. 7, 1992, p. 25.

Emerging Telecommuting Services 9

The previous six chapters identified the communications needs of the individual telecommuters (marketing, technical, remote agents, and administrative), as well as those of RTCs. These chapters also proposed specific networking solutions to meet the needs of the individual telecommuters (ISDN, POTS, DOV, and ISDN/frame relay) and to meet the needs of the RTCs (private lines, SMDS, frame relay, ATM, and switched/dialup services). This chapter explores three types of emerging solutions: (1) an emerging RTC networking solution—native mode LAN interconnection service (NMLIS), (2) emerging networking solutions that can address the communications requirements of telecommuters—asymmetric digital subscriber line (ADSL) and a hybrid cable POTS solution, and (3) an emerging application solution-integrated telephony services.

9.1 NATIVE MODE LAN INTERCONNECTION SERVICE

Native mode LAN interconnection service (NMLIS) is a family of LAN extension and channel extension services designed to support connectivity between the same types of LANs (Ethernet, token ring and fiber distributed data interface—FDDI) and high-performance computer systems across metropolitan area networks and wide area networks [1]. These services operate at native LAN speeds (4, 10, and 100 Mbps). NMLIS can support the networking requirements of RTCs, providing RTCs with a substitute to the following services: frame relay, SMDS, and private T1 services.

NMLIS is currently offered by several LECs [2], such as Bell Atlantic (FDDI network services and wire speed LAN interconnection service), BellSouth (Ethernet interconnection service, token ring interconnection, and FDDI interconnection service), and U S WEST (transparent LAN service) [3]. NMLIS is also offered by several AAPs. For example, Metropolitan Fiber Systems

(MFS) offers high-speed LAN interconnection (HLI) service [4]. These services have several common characteristics, including the following.

- They are based on a dual counter-rotating fiber optic ring infrastructure.
- These rings are initially deployed in urban areas as MANs and then extended over WANs.
- They can only interconnect the same LANs across MANs. For example, NMLIS may provide an Ethernet-to-Ethernet connection or a token ring-to-token ring connection.
- *Preconfigured connectivity* is required to implement the service; that is, the service does not provide any-to-any connectivity. Instead NMLIS is appropriate for customers that have requirements for interconnecting a few nodes in a MAN with predictable destinations.

While the NMLIS offerings of the various communications service providers have some similar characteristics, they also have some different characteristics. There are two major versions of NMLIS: (1) a dedicated version, and (2) a virtual version. Each communications service provider may offer one or both versions. Each of these versions is described next.

9.1.1 The Dedicated Version

The dedicated version is based on time division multiplexing (TDM) technology. The dedicated version means that bandwidth equal to the native LAN rates of 4, 10, 16 Mbps and/or channel extension rates of 3 or 4.5 Mbps is dedicated to a customer. This version is appropriate for RTC participants if they have high throughput requirements but cannot tolerate delay. Several communications services providers offers a dedicated version of the service. An example of a communication service provider that offers this service is MFS, which plans to offer this service in over 24 urban areas across the United States.

As Figure 9.1 shows, RTC participants occupying a floor of a building may access a fiber-optic ring supporting NMLIS and communicate with other locations on the ring through a concentrator, which is located in the basement of the building. Only the communications service provider has access to the concentrators. The communications service provider is responsible for procuring, installing, and maintaining the concentrator. An example of these concentrators is FiberMux's Magnum FX 4400 and FX 8800. This equipment performs *multiplexing and bridging* functions. FiberMux's Magnum FX 4400 is a 45-Mbps TDM that provides connectivity to 1 to 4 Ethernet segments, 1 to 8 token rings, 1 to 16 T1 circuits, and one to 32 V.35 channels. FX 8800 is a 100-Mbps TDM that provides connectivity to 1 to 8 Ethernet segments, 1 to 16 token rings, 1 to 32 T1 circuits, 1 to 64 V.35 channels, or any mix of these signals over a dual counter-rotating ring.

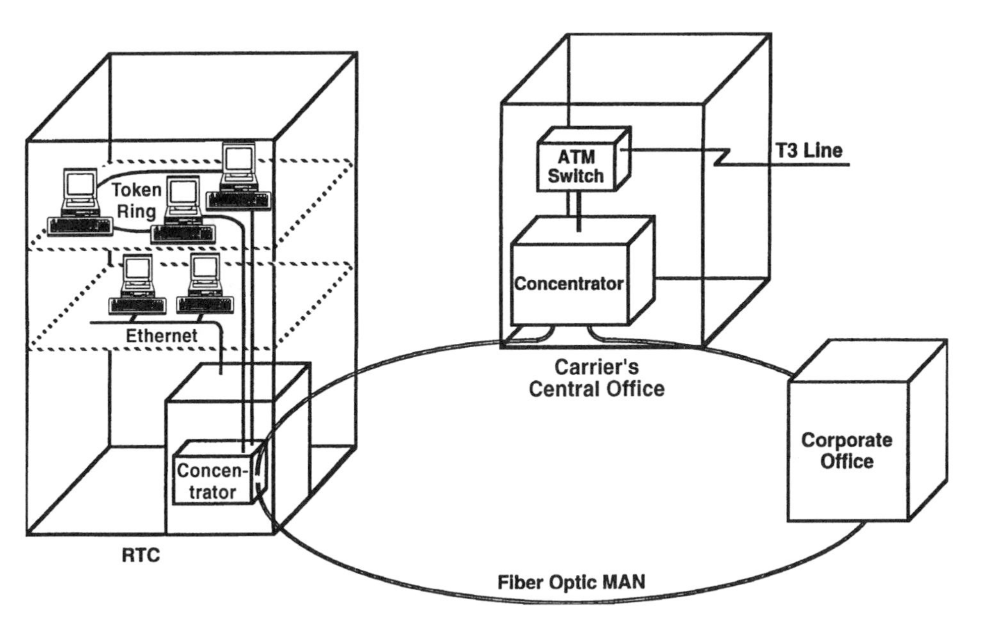

Figure 9.1 RTC/corporate office links: the native mode LAN interconnection solution.

Several MANs supporting the dedicated version of NMLIS can be interconnected over a WAN by relying on several technologies, including SMDS, frame relay, and ATM. Figure 9.1 shows the NMLIS service extended to the WAN by relying on an ATM switch and T3 lines. These T3 lines are leased from an interexchange carrier. ATM provides the NMLIS provider with the ability to format data into cells, combining traffic from different users on the same backbone and providing the customers with cost savings over leased T3 lines. ATM also provides the customer with data security through its virtual circuit connection capabilities. ATM's role could be expanded beyond the interconnection of MANs across a WAN by replacing the concentrators located in customer buildings with ATM switches.

9.1.2 The Virtual Version of NMLIS

The virtual version of NMLIS is based on FDDI technology. The virtual version allows users to interface to the service at native LAN rates of 4, 10, 16, or 100 Mbps and/or speed-reduced channel extension rates. Unlike the dedicated version of the NMLIS, users are not guaranteed the native LAN speed. This is because users share the FDDI fiber-optic backbone. Consequently, they may have to contend for the bandwidth available on the backbone. The virtual version of NMLIS is appropriate for RTC participants if they have high throughput requirements but are willing to tolerate delay in exchange for cost savings. Data privacy on the shared ring is provided by an IP-based encapsulation protocol. An example of a communication service provider that offers an FDDI-based virtual service is Bell Atlantic, which provides FDDI network services (FNS) in several urban areas.

In the case of a virtual version of the NMLIS, RTC participants occupying a floor of a building may access a fiber-optic ring supporting NMLIS and communicate with other locations on the ring through an FDDI bridge or router, which is located in the basement of the building. As in the case of concentrators, these bridges or routers are located in the basement of the building in a sealed room. Several manufacturers offer FDDI bridges or routers. An example of these bridges or routers is Fibercom's Ring Master™ 7200 FDDI bridge. This bridge operates at the MAC layer and can support the interconnection of four LANs per bridge (e.g., IEEE 802.x)

NMLIS offers users several benefits, including the following.

1. *High transmission speeds:* NMLIS provides RTCs with higher transmission speeds than frame relay services and the same range of speeds associated with SMDS.
2. *Support for bursty data traffic*: NMLIS is well-suited for RTCs that have extensive needs for bursty data traffic communications at high bandwidth levels. NMLIS provides the IS manager with a choice of various native LAN speeds, including 4, 10, 16, and 100 Mbps.
3. *User equipment management:* NMLIS reduces the network management cost of users in comparison with other fast-packet services (e.g., frame relay and SMDS) because it enables them to outsource the procurement, design, installation, maintenance, operations, and software upgrade of concentrators (or the FDDI-based routers) located in the basement of their buildings.
4. *Economics*: NMLIS *may* provide an economic alternative to frame relay service and SMDS because while prices of frame relay service and SMDS are fixed, the prices of NMLIS varies. Several factors may determine the leasing cost of NMLIS, including the number of connections, the speed requirements associated with these connections, and the distance between the locations.
5. *Technology independence*: NMLIS is a technology-independent service. It can be implemented over FDDI or TDM technology. Buyers of the service do not have to learn about the technology associated with the service, reducing the time they need to allocate to learning about the service.

While NMLIS can provide an RTC with several benefits as explained above, it has several shortcomings, including the following.

1. *Lack of support for isochronous traffic*: NMLIS only supports asynchronous packet data traffic. It cannot support POTS voice, signaling, and features (e.g., CLASS), and real-time video applications. Consequently, if RTCs are populated predominantly by marketing telecommuters who, as indicated earlier have video communications applications, frame relay service cannot provide a total solution to the RTC. Consequently, NMLIS is not appropriate unless it is complemented by other solutions.
2. *Preconfigured connectivity*: NMLIS is only appropriate for those users that have requirements to interconnect a few nodes with predetermined destinations. It is not, however, appropriate for any-to-any connectivity, as is supported by SMDS.
3. *Availability:* The availability of NMLIS is limited to several select urban areas. This is attributed to the fact that NMLIS is an emerging service. LECs and AAPs will likely expand the availability of the service in the coming years.
4. *Lack of extension over WANs:* NMLIS predominantly exists in urban areas. The support by communications service providers for interconnecting NMLIS across WANs remains limited.

9.2 EMERGING TELECOMMUTER/CORPORATE OFFICE NETWORKING SOLUTIONS

Two of the solutions that are emerging in support of the telecommuter's communications needs are a hybrid cable TV/POTS solution and an asymmetric digital subscriber line (ADSL). The cable TV/POTS solution complements the existing dialup/modem solution, while ADSL represents a future enhancement to the existing telephone network infrastructure, providing the telecommuter with higher networking performance. Each of these solutions is examined next.

9.2.1 The Hybrid Cable TV/POTS Solution

This solution is being offered on a limited basis by a small number of cable TV companies (e.g., TCI), as part of their efforts to expand their markets beyond residential customers. For example, TCI envisions the future as the connection between cable and the public telephone network. TCI uses cable as a downstream, one-way link to homes, with the telephone network as the return path. Telecommuters, who must be cable subscribers, are connected to the head-end via a modem that is connected to their PCs or through a card that is loaded into the machines. Data is received via a standard cable channel and is transmitted as any basic or pay channel without affecting any other cable service [5].

On the corporate site, the company's computer is also connected to the cable system's head end via a modem. Once a corporate computer is connected to the cable's head end, access to the corporate computer is available to any telecommuter with little incremental cost to the cable system. Data travels to the home over the cable system at speeds of 1.5–10 Mbps, enabling the telecommuters who are participating in the program to download large files much faster than by relying on a dialup line. The telephone network is used as the return path. The cable TV company is responsible for providing necessary security access.

This solution offers the telecommuter several benefits over the dialup/modem solution, including (1) higher bandwidth in one direction, enabling the telecommuter to download large technical files rapidly; (2) the ability to receive and make phone calls during data communications sessions; and (3) the ability to simultaneously handle data and cable TV services on the same line.

This solution, however, has several limitations, including (1) it does not provide the telecommuter with access to advanced voice management capabilities; (2) it is not widely available, at least for the moment; (3) it does not enable the telecommuter to communicate with those who are noncable TV users; (4) it does not provide loop backup powering during a local power outage; and (5) the costs of this solution are still unclear because the technology and associated services are immature. In addition, some rewiring of telecommuter premises will be needed to install this service. However, many cable TV companies are planning to enter the telephone business. For example, in August 1994, a consortium of six major cable TV companies—TCI, Time Warner Cable, Continental Cablevision, Comcast, Cox Communications, and Viacom International—representing about 26 million subscribers announced a $2 billion plan to offer a variety of local telephone services. The investment seeks to deploy network interface cards capable of supporting telephone and video connections in the home [8].

9.2.2 Asymmetric Digital Subscriber Line

Until recently, there were a few viable high-performance copper-based solutions available to the RBOCs in addressing the needs of residential customers (including the telecommuters) for voice, data, and video services. Recent advances in digital signal processing (DSP) have resulted in large increases in the amount of data that the telephone companies can transmit over copper wire. These capabilities have been incorporated in a new technology—asymmetric digital subscriber line [8].

ADSL was first proposed by Bellcore in 1989. Since then, ADSL has evolved significantly and has gained the support of a number of RBOCs. ADSL is currently undergoing a standardization process through the T1E1.4 subcommittee of the Alliance for Telecommunications Industry Solutions. This subcommittee defined the following capabilities of an ADSL system.

1. *One-way high bandwidth (downstream channel)* [6]: From the CO to the telecommuter, ADSL may provide the telecommuter with data rates of 1.5 Mbps up to 6 Mbps. This bandwidth may be split among multiple channels. For example, 6 Mbps may be split among four channels, each capable of supporting a VCR-quality video signal (MPEG1 compression). This bandwidth may also be split among two channels, each capable of supporting a sports-quality (MPEG2) real-time video signal.
2. *One bidirectional ISDN H0 channel operating at 384 Kbps* [7]: This full duplex channel, which is compatible with dialup circuit-switched services, including those at fractional T1 rates, can provide a corporation with the ability to extend personal videoconferencing to the telecommuter's home. This would be particularly useful to a marketing telecommuter who, as mentioned in Chapter 5, can benefit from videoconferencing. This channel can also support the telecommuter's needs to access corporate LANs through frame relay.
3. *One bidirectional 160 Kbps ISDN basic rate channel*: This contains two B channels, a D channel, and associated ISDN operations channels.
4. *A control channel:* This channel enables telecommuters and other residential customers to control service delivery, including the ability to fast-forward, reverse, search, and pause features for movies by relying on a remote control.
5. *An embedded operations channel*: This is for internal systems maintenance, audits, and surveillance.
6. *Passive coupling of ADSL to basic phone service*: ADSL-based services are delivered to telecommuters and other residential customers over the same copper pair that support basic telephone services. The coupling of ADSL and the basic phone line, however, is passive. This means that if the ADSL system fails, the telecommuter can still place and receive phone calls.

ADSL can provide telecommuters with these capabilities by relying on discrete multitone technology (DMT) [8]. This technology maximizes throughput over a copper pair by dividing available bandwidth into 256 subchannels and allocating incoming data to each subchannel according to its ability to send data. If any subchannel cannot send data, it is shut off. The other working subchannels are modulated so that they are able to carry from 1 to 11 bits per symbol based on learned channel characteristics.

In a DMT-based ADSL system, a copper pair may be divided into three channels: (1) POTS occupies the baseband and is split from the digital data channels by passive filters or other means; (2) an upstream digital channel carries data in various configurations up to 384 Kbps; and (3) a downstream channel occupies the remaining bandwidth, and carries up to 6 Mbps of asymmetric bandwidth as well as the downstream portion of bidirectional channels—up to 576 Kbps (Figure 9.2).

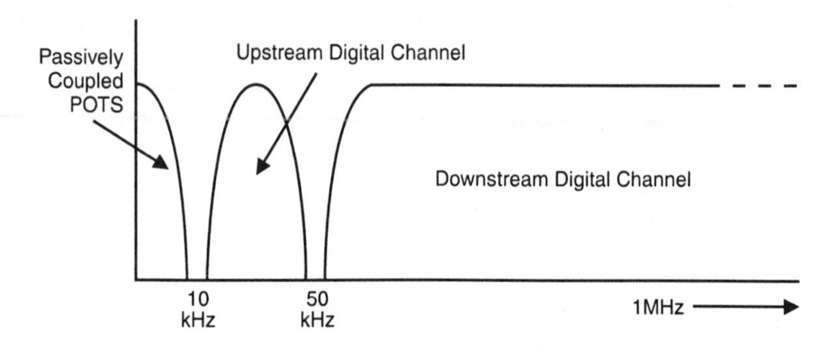

Figure 9.2 ADSL frequency spectrum.

ADSL is currently being evaluated by a number of RBOCs. The most active of the RBOCs in the field of ADSL is Bell Atlantic, which has won a patent for its approach to ADSL. The patent describes a system that provides a digital video signal from a video information provider to one or more subscriber homes. The subscriber orders programming by either relying on a standard telephone set or by using a dedicated control device over a packet link [8].

Following a service order, Bell Atlantic establishes a link between the information provider and the CO serving the subscriber. Connectivity between the CO and the subscriber is established using ADSL. Interface units use frequency multiplexing technology to deliver digital video information with voice information to the subscriber and support transmission of a reverse control channel. These interfaces allow baseband signaling and audio for conventional telephone service.

In addition to patenting ADSL, Bell Atlantic conducted a trial in 1993 [9]. Bell Atlantic used an IBM RS/6000 workstation and software it developed inhouse for the video server. Bell Atlantic also used prototype equipment from Northern Telecom, AT&T Network Systems and Westell, Inc. (Oswego, IL). The tests provided 1.5 Mbps to homes which were located at a distance of up to 18 kft from the CO, 3 Mbps to homes within 10,000 to 12,000 feet from the CO, and 6 Mbps to homes within 6,000 feet from the CO.

ADSL can potentially offer telecommuters several benefits, including (1) the ability to support POTS voice, signaling, and CLASS features; (2) the ability to deliver POTS during local power outage; (3) simultaneous access to voice, data, and video services over a single line (although video is only one way); (4) higher bandwidth than ISDN BRI in one direction, supporting the requirements of the more demanding telecommuters (e.g., technical telecommuters) to download files from the corporate office; and (5) the ability to receive voice services without rewiring telecommuters' homes.

ADSL, however, cannot be ordered today because the technology to be used to transport this capability is not yet finalized. Also, a number of RBOCs are not actively planning to deploy the technology, but are pursuing instead hybrid fiber coax (HFC) solutions.

ADSL may not be an appropriate networking solution for several years. This can be attributed to several reasons: (1) ADSL development work will take several years, and (2) the economics of this technology are not likely to be favorable in the next few years to support telecommuting solutions.

ADSL may not be widely available even after several years. This is because (1) ADSL technology is subject to distance limitations, (2) ADSL requires good-quality copper facilities while the quality of installed based of copper facilities vary widely from one region to another, and (3) as noted, several RBOCs are exploring other alternatives to ADSL for video delivery, such as fiber/coax solutions.

9.3 INTEGRATED TELEPHONY SERVICES FOR TELECOMMUTERS

Integrated telephony services differ from the last three solutions described in this chapter in one fundamental way: NMLIS, ADSL, and cable TV/POTS are network-level solutions. In contrast, integrated telephony service is an application-level solution that runs in the user's PC. (It can use any of the communications services described in this book. Initially the target is POTS.)

Much has been said about the convergence of computers and telephony; however, where it would be most obvious to the casual observer, namely, at the desktop, the PC and the telephone by and large seem to remain separate, distinct, and unfriendly. That may soon change with the agreement and promulgation of telephone-specific application programming interfaces (APIs) and products supporting these APIs. This section discusses what APIs are and how they can foster the symbiotic relationship that can open the door for productivity enhancements at the desktop and elsewhere. It focuses on the Intel/Microsoft API, with a short discussion on IBM's CallPath services architecture.

The telephony services made available by these APIs can be of value to telecommuters seeking to increase their productivity. The business community is interested in technology for productivity enhancements, not just for sheer elegance. Telephony API is one such technology.

9.3.1 Application Programming Interface

Application programming interfaces enable application developers to build applications without worrying about the detailed environment below the "line" defined by the API. Typically this line resides just above the application layer of the open system interconnection reference model. The programmer utilizes generic descriptors of actions that are desired below the line; these descriptors are then properly mapped by the underlying platform. The platform can consist of different hardware families; hence, the actual translation of the programmer is platform-dependent.

Recently an API has been defined for telephonic functions under Microsoft Windows (Version 3.1 or higher). This Windows telephone API (TAPI) provides services that enable an application developer to add elements of telephony to Windows-based applications. The Windows telephony services are provided as a WOSA (Windows open services architecture) component. It consists of both an API used by the applications and a service provider interface (SPI) implemented by service providers. Only the API is described in this section.

With the realization of telephony services on a PC, users can perform time-saving and value-enhancing tasks in an easy manner. For example, one can add callers to a conference call by dragging names out of directories and dropping them on telephone icons, or there may be a personal information manager that allows users to match the incoming caller ID information with other information such as name, records of previous conversations, caller's hobbies, and "spouse and kids" data. These applications can also offer visual call control: voice-mail, e-mail, and fax integration; and desktop audio and video conferencing.

9.3.2 Use of Windows Open Services Architecture

What is WOSA? WOSA provides a single-level interface for connecting front-end PC applications with back-end services and utilities. As a general-purpose architecture, it goes beyond the telephony environment. With WOSA, application developers and users need not worry about conversing with numerous services, each with its own protocols and interfaces, because making these connections is the task of the operating system, not of individual applications. WOSA provides an extensible framework in which Windows-based applications can seamlessly access information and resources in a distributed environment. WOSA accomplishes this by making a common set of APIs available to all applications.

The front-end application and back-end service need not know each other's language in order to communicate as long as they both know how to talk to the WOSA interface. As a result, WOSA allows applications developers and vendors of back-end services to mix applications and services to built solutions that shield programmers and users from the underlying complexity of the system.

WOSA defines an abstraction layer to heterogeneous computing resources through a set of (WOSA) APIs. Because this set of APIs is extensible, new services and their corresponding APIs can be added as needed. Applications written to the WOSA APIs have access not only to all the various computing environment supported at this juncture, but also to additional environments as they become available. The key advantage is that applications do not have to be modified to obtain this support.

Each service by WOSA also has a set of interfaces that service-provider vendors use. In order to provide transparent access for applications, each implementation of a particular WOSA service needs to support the function defined by its service-provider interface. WOSA uses a Windows dynamic-link library (DLL) that allows software components to be linked at runtime. An application needs to know only the definition of the interface, not its implementation. Figure 9.3 shows the WOSA environment.

Telephony services under Windows, discussed next, follow the WOSA model. This implies the need to define a telephony API that is the application programmer access to the telephony services. To be of value, an API must be "standardized" so that all developers can use it reliably. In addition, there is a telephony SPI that is implemented by telephone service vendors, and a telephony DLL that is (to be) part of the Windows operating system. The telephony API may be used simultaneously by multiple applications.

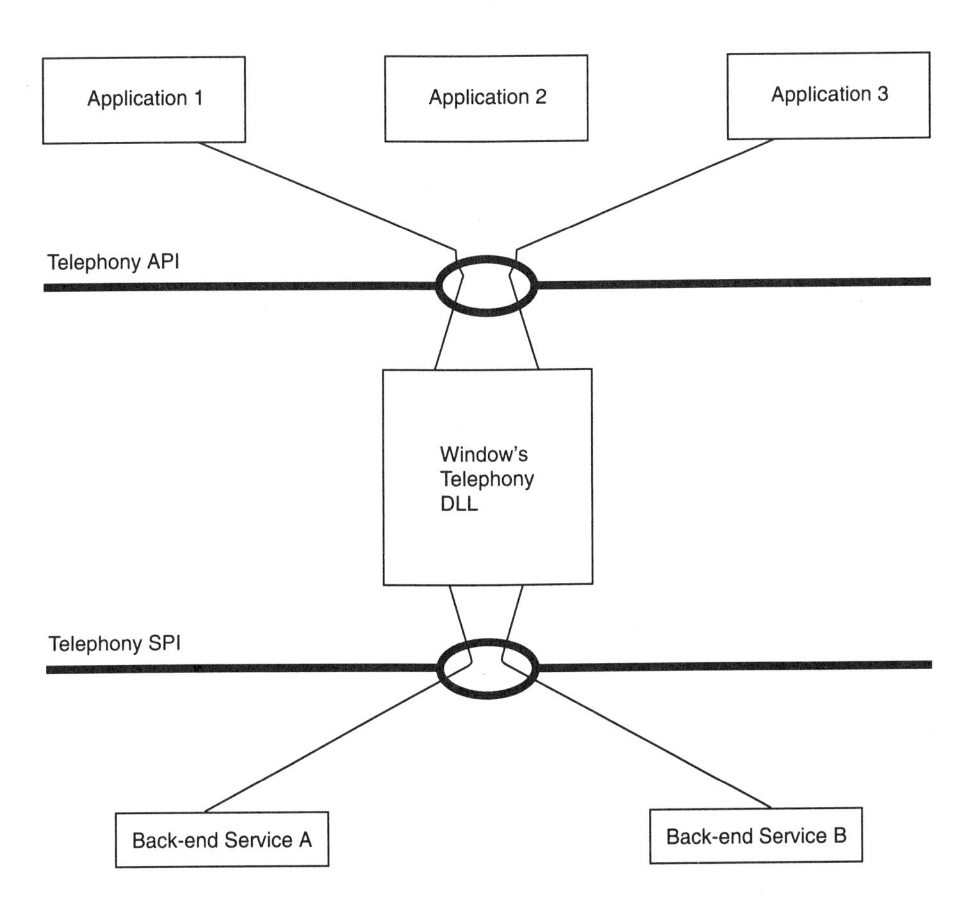

Figure 9.3 WOSA.

9.3.3 Telephony Services Under Windows

In the past, linking an application, or more generally a computer to a telephone system, has not proven to be an easy task. Many PBXs, key systems, and telephone company switches use proprietary protocols at the upper as well as at the lower (physical) layers. The focus of the telephony API is to provide "personal telephony" to the Windows platform. This gives developers of applications such as word processing, spreadsheets, and so forth a means to enable users to control real-time communications directly in the context of their work. Features such as "intelligent" answering, coordinated voice/data transfer, preview dialing, and predictive dialing become available.

The API is independent of both the telephone network that performs the actual switching (Centrex, PBX, ISDN-based, cellular, etc.) and the connectivity mechanism used to link the PC to the network providing the telephone service.

Telephony services can be partitioned into "simple telephony services" and "full telephony services." Simple telephony allows telephony-enabled applications to easily incorporate call establishment from within these applications without these applications needing to become aware of the details of the full telephony services.

The Windows telephone API defines the interface that applications use to access telephony functions in Windows. The API is a collection of C language function definitions, message definitions, types, and data structure definitions, along with enough English verbiage to describe their meanings completely and unambiguously. (Applications are any software that uses the API; it need not be a user-level program, but can also be a DLL that uses the API and provides a higher level of service via its interface.) A switch is a piece of equipment capable of establishing telephone calls. Within the context of the API, a switch can be a PBX, a key system, or a central office switch.

Complete call control is only possible through the use of the full telephony services. Applications access the full telephony API services using a first-party call control model. This means that the applications controls telephone calls as if it is an endpoint of the call. The applications can make calls, be notified about inbound calls; answer inbound calls; invoke switch features such as hold, transfer, conference, pickup, park, and so forth; and detect and generate DTMF for signaling with remote equipment.

The fact that the API presents a first-party call control does not restrict its use to only first-party telephony environments. The Windows telephony API can be used for third-party call control.

The API provides an abstraction of telephony services that is independent of the underlying telephone network and the configuration used to connect the PC to the switch and phone set. The API provides an abstraction of the PC connections to the switch or network and the phone set. The connection may be realized in a variety of arrangements including pure client-based wire or wireless connections or client/server configurations using a local area network.

The API allows the application to control the call progress. Call progress refers to setting up a telephone call. There are several phases to this process. After taking the phone off hook, the switch returns a dialtone to indicate that a number can be dialed. Hearing the dial tone, the user dials the desired number. When the call reaches the destination phone, the caller will either receive a busy indication or a ringback indication that the dialed party is being altered.

The telephony API by itself is not concerned with providing access to the information exchanged over a call, but with the call control (the telephone API can work in conjunction with other Windows services such as the Windows multimedia wave audio, media control interface (MCI), or auto-fax APIs to provide access to the information on a call).

The telephony API defines three levels of service. The lowest level of service is called basic telephony and provides a guaranteed set of functions that corresponds to POTS: make calls and receive calls. The next service level is supplementary telephony service, providing switch features such as hold, transfer, and so forth. All supplementary services are optional. Finally, there is the extended telephony level. This API level provides API extension mechanisms that enable application developers to access service-provider-specific functions not directly defined by the telephony API.

The specification, called "Windows Telephony API" (May 1993—Preliminary) has been developed jointly by Intel and Microsoft. Microsoft provides the system resources while Intel plans to offer PC boards, cameras, handsets, and other add-on products for videoconferencing. In addition to Microsoft and Intel, early supporters included Novell Inc., Sun Microsystems (SunConnect), and Synoptics Communications. The specification has been endorsed by over 40 vendors, including AT&T, Northern Telecom, Octel Communications, and Digital Equipment Corporation.

9.3.4 Some TAPI Functionality

The API specification entails function messages and parameters. The names of functions consist of descriptive words, such as "lineMakeCall" (function that establishes a phone call) "lineAnswerCall," "tapiRequestMakeCall," and so forth. The prefix "tapi" indicates that the function belongs to the simple telephony API. The prefix "line" or "phone" indicates that the function belongs to the full telephony API, "line"-prefixed operations work on the connection to the network, and "phone"-prefixed operations work on a telephone set (this is a phone-type device that is somehow associated with the PC running the application and controlled by the API).

Messages can be sent to the application's window or to an application's callback function. Messages always begin with the prefix "TAPI," "LINE," or "PHONE" ("TAPI"-prefixed messages are sent to the application's window; "LINE" and "PHONE"-prefixed messages describe the parameters passed to the callback function).

The Windows telephony service provider is the conglomerate of software code (DLL, device drivers, firmware) and hardware (add-on hardware, phone set, switch, and network) that jointly implement the telephony API.

Simple telephony provides telephony-enabled applications with an easy mechanism for making phone calls without requiring the developer to become telephone literate. Two services are provided:

1. Request the establishment of a voice call between the user and a remote party identified by the party's telephone number. The request is made to the telephone API, which passes the request on to an application that has registered with the API as a recipient of such requests. This is usually the user's call control application. A basic sample call control application is provided as part of the developer's toolkit. After the simple telephony application has made the request, the call is totally controlled from the call control application; the simple telephony application is enabled to manage the call. Applications such as word processors, spreadsheets, schedulers, personal information managers, and so forth can become telephone aware via the tapiRequestMakeCall function. Since the complexity of the full telephony API and all user-interface aspects of telephony are handled by the call control application, the telephony-enabled applications need not be modified in any substantial way.
2. Request the establishment of a media call with the intent of accessing the corresponding media stream (voice, fax, data, etc.) via the appropriate media API. As in the previous case, the request is made to the telephony API, which passes the request on to an application that has registered with the API as a recipient of such requests. After the simple telephony application has made the request via the tapiRequestMediaCall, the application is informed about the success or failure of the call establishment via the TAPI_REPLY Windows message. Once the call is established, the media stream can become active and is under control of the application. The only other operation on the call is tapiRequestDrop, which requests that the call be dropped. Applications can use this service if they are primarily interested in managing media streams, but want to leave the details and user interface aspects of call setup to a call control application.

9.3.5 Examples of Arrangements

In TAPI, a "phone" is a device that behaves as a telephone set. This is usually (although not necessarily) the phone already on the user's desk located "next to" the PC. Phones need not be physically connected to PC. A LAN-based server with appropriate access to the switch may provide this logical connection.

The telephony API treats the phone and the line as separate devices that can be independently controlled by the application. Therefore, a phone is a device that implements the phone behavior defined by this API as the set of functions and message for phones.

As an example of connectivity supported by TAPI (Figure 9.4), the PC can be connected to the switch through the telephone set. Such sets typically connect to the PC via a serial port. The functionality supported by the API is mapped onto commands sent over the serial connection to the phone (e.g., the Hayes Modem command set).

As another example (Figure 9.5), the PC can use an add-on card that has telephony connections to both the switch and the set.

Another implementation (e.g., Figure 9.6) may entail a LAN-based server with multiple connections to the switch. API operations invoked at any of the

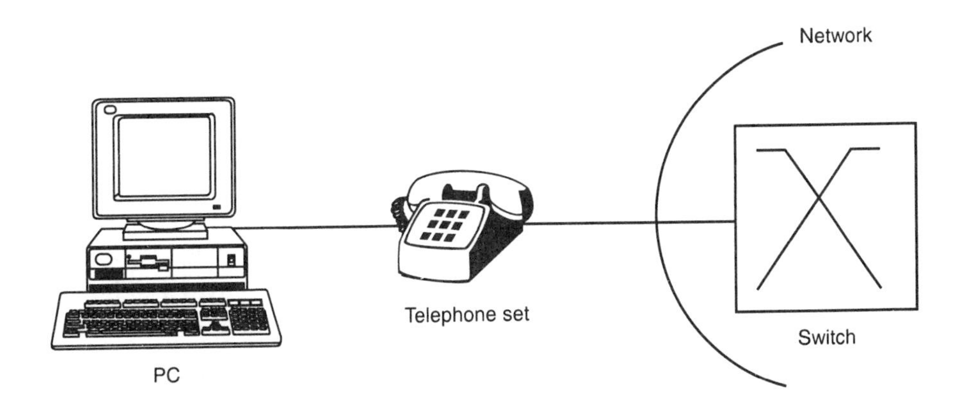

Figure 9.4 PC connected through the telephone set.

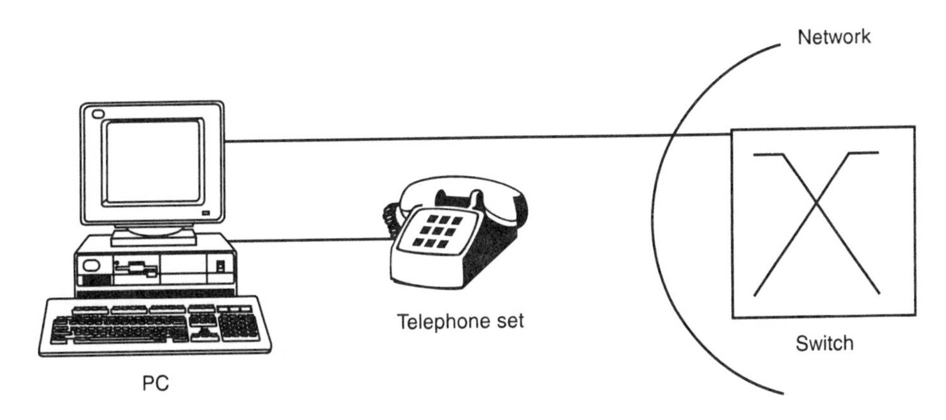

Figure 9.5 Direct connection with the switch.

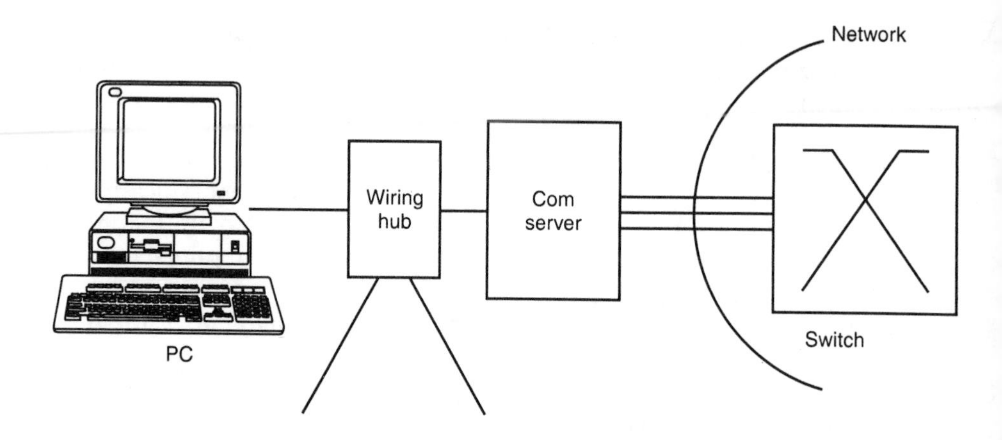

Figure 9.6 LAN-based server.

client PCs are forwarded over the LAN to the server. The server uses third-party call control to implement the client's call control requests.

As yet another alternative, the server is connected to the switch over a switch-to-host link. API operations invoked at any of the client PCs are forwarded over the LAN to the host. The host uses third party switch-to-host protocols to implement the client's call control requests.

9.3.6 CallPath Services Architecture

In the early 1990s, IBM has introduced an "open" architecture known as CallPath services architecture (Figure 9.7). From the end-user's perspective, CallPath, when implemented on an IBM system provides an API that masks the differences between telephone systems. Communication between the IBM systems (e.g., PS/2, AS/400, and S/390) with CallPath services API and a telephone system that links to the IBM computer using this mechanism, takes place through an application receiving messages from the telephone system defined by the API. This predefined format frees the user from the peculiarities of the actual link between the two systems. Similarly, a request for service can be presented by the application at the API and it is sent to the telephone system without the application's becoming concerned with the link specifics.

The CallPath services application API is available on the PS/2 (both the OS/2 and DOS for Windows operating systems); it is available on departmental systems (AS/400) and in the mainframe environment (MVS and VSE). Telephone equipment platforms that have announced CallPath implementations include AT&T, NEC, ROLM, Northern Telecom, and Siemens.

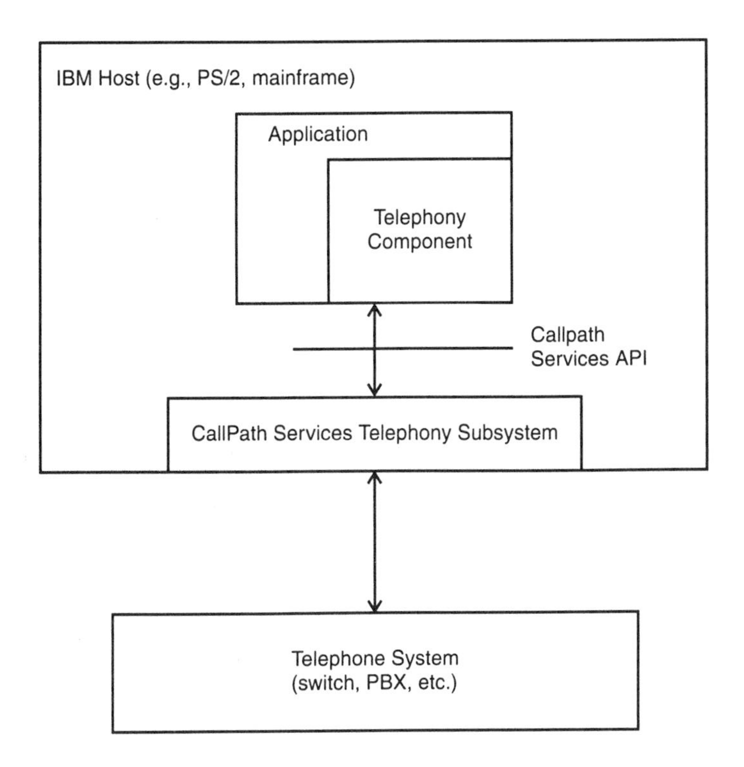

Figure 9.7 CallPath.

9.4 CONCLUSIONS

This chapter explored four types of emerging services and technologies:

- The first service, NMLIS, provides RTCs with an alternative to SMDS and frame relay services.
- The second service, cable TV/POTS, provides a complement to the existing dialup/modem solution, providing the telecommuter with higher performance in one direction.
- The third service, ADSL, is a higher performance service, providing telecommuters with a platform in support of data, voice, and one-way video applications. ADSL is a promising technology that can support ISDN. This service, however, will not be available for many years. Ultimately, the success of this service will depend on its availability and cost structure.
- The fourth service described in this chapter has the potential of providing telecommuters with telephony services on a PC, enabling them to perform time-saving and value-enhancing tasks in an easy manner.

References

[1] Eldib, Osman, "A Comparison and Description of Broadband Services, *Supercomm Conference*, May 2, 1994.
[2] "Local Carriers Getting their Hands around LANs," *Communications Week*, March 1, 1993, p. 40.
[3] Wallace, Bob, "U S West Unit Introduces LAN Interconnection Service," *Network World*, Aug. 24, 1992, p. 19.
[4] Fahey, Michael, "Company Builds National ATM Fiber Network," *Lightwave, The Journal of Fiber Optics*, Nov. 1992.
[5] "TCI Provides Stay-at-Home and Go-To-Work Service," *Metropolitan Network News*, April 25, 1992, p. 7.
[6] Fleming, Stephen, and Michael B. McLaughlin, "ADSL: The On-Ramp to the Information Highway," *Telephony*, July 12, 1993, p. 20.
[7] "Showcase Applications," Northern Telecom's ADSL System, *Supercomm '94*.
[8] Minoli, D., *Video Dialtone Technology, Approaches, and Services—Digital Video Over ASDL, HFC, FTTC, and ATM*, New York, NY: McGraw-Hill, 1995.
[9] "HDSL and ADSL: Steroids for the Local Loop," *Business Communications Review*.

The Network Service Providers 10

This chapter focuses on the supply side of the telecommuting industry. The purpose of this chapter is to provide existing and potential developers, implementers, and participants in a telecommuting program with a guide to the current and emerging providers of the communication solutions that have been examined throughout this book.

The chapter first explores the current network service providers and their communications services. The chapter then discusses the potential providers of communications solutions to telecommuters. This chapter concludes with an identification of the telecommuting business opportunities available for network service providers in the telecommuting markets. Although some of this information is time-dependent, all of the network service providers and many—if not all—of the services are expected to remain available in the future.

10.1 CURRENT NETWORK SERVICE PROVIDERS

The telecommuting market is currently supported by two major groups of communication service providers: the local exchange carriers (LEC) and the interexchange carriers (IXCs). These two industries are the most active groups of communications service providers in the telecommuting market. Recognizing the telecommuting market as a separate market segment distinct from both the residential and the business segment, they have taken major steps to establish their presence in this market.

10.2 THE LOCAL EXCHANGE CARRIERS

The LECs have been among the most active telephone companies in the telecommuting market. The services that the LECs offer can support most of the requirements of the different types of telecommuters as well as those of the RTCs. The regional Bell operating companies (RBOC) are the largest LECs. There are several similarities among the services that RBOCs offer, including the following.

1. *All RBOCs support the deployment of ISDN,* although as mentioned in the text, they differ in terms of the speed of deploying ISDN.
2. *All of the RBOCs support the deployment at least one broadband data service,* such as SMDS or frame relay.
3. *All of the RBOCs support ATM* as a platform in support of FRS and SMDS, as well as ATM-based services such as ATM cell relay.
4. *All of the RBOCs are experimenting* with various approaches to providing broadband video distribution services to the residential market (e.g., ADSL, hybrid fiber/coax solutions).

The following is a brief description of the services offered by some of the major LECs.[1]

10.2.1 Ameritech

Ameritech offers telecommuters a number of telecommuting services, including a five-state edition of a bimonthly newsletter. Ameritech has also been actively deploying ISDN [1]. According to industry forecasts, more than 75% of Ameritech lines are planned to have ISDN access by 1995 [2]. Ameritech also planned to offer multirate ISDN, a variation on the ISDN PRI, which can operate at speeds in multiples of 64 Kbps. In addition to ISDN, Ameritech has made a major commitment ($4.4 billion) to hybrid fiber coax video services, with the target of 6 million customers by 1999.

To support the needs of business customers, Ameritech deployed ATM switches in Chicago and Milwaukee in February 1993. Ameritech's initial offerings on its ATM platform were expected to emphasize SMDS and to include a full range of bandwidths from 64 Kbps to 384 Kbps [3]. In February 1993, Ameritech introduced the Ameritech LAN interconnect service (ALIS) after successfully completing field trials with a number of customers. In addition, Ameritech introduced a 384 Kbps fractional T1 service throughout its five-state region [4].

---

1. This information is based on published reports; although it is believed to be correct, it has not been verified with the carriers.

10.2.2 Bell Atlantic

Bell Atlantic demonstrated its support for the telecommuting concept by implementing a telecommuting program internally.[2] In the fall of 1992, Bell Atlantic gave 16,000 managers the option of staying at home to work [5]. Bell Atlantic planned to open the program to all management employees who passed an assessment to determine whether the job of the applicant to the telecommuting program was suitable for telecommuting. This large program followed two pilot trials. The first pilot program, which began in 1991, was intended to determine the degree of success of alternative work arrangements for employees and their bosses and to measure the impact on work productivity. In this pilot, 27% of the participants received higher performance ratings over the previous year when they were not telecommuting. The second program started in February 1992. Bell Atlantic provided each of the participants in the program with a second phone line at home, voice mail, e-mail, and three-way calling. Bell Atlantic planned to showcase their own programs with their customers to help them learn from Bell Atlantic's experience.

Bell Atlantic also supports telecommuters and other residential customers by increasingly offering ISDN. In September 1993, Bell Atlantic started offering individual-line ISDN to small businesses in Virginia, West Virginia, Maryland, and Washington DC. In 1994, ISDN BRI was expected to become region-wide and available to residences. In addition to ISDN, Bell Atlantic tested the ADSL technology in May 1993 [6] in a trial that involved 70 employees living in Northern Virginia. They also have an interest in hybrid fiber coax cable TV service.

To meet the needs of the business customers, Bell Atlantic introduced SMDS and FRS [1] in addition to an FDDI-based version of NMLIS.

10.2.3 BellSouth

To serve the needs of the telecommuting market, BellSouth offers a comprehensive guide to help corporate customers through the process of establishing a telecommuting program. This guide is written specifically for human resources personnel who are responsible for managing corporate telecommuting programs. The guide is written as an action plan and covers the technological as well as the human resource issues.

BellSouth also deployed ISDN extensively to serve the needs of the residential and the telecommuting markets [7]. According to industry forecasts, almost 60% of BellSouth lines were planned to have ISDN access by 1995. BellSouth tariffed two versions of ISDN throughout its nine state territory: ESSX ISDN (basic rate/Centrex) and MegaLink ISDN (PRI). In 1993, BellSouth

---

2. This information is based on published reports; although it is believed to be correct, it has not been verified with the carriers.

also trialed an individual line ISDN [8]. In 1993, BellSouth also introduced a tariff allowing a business line to the home at residential prices.

To meet the requirements of the business market, BellSouth trialed NMLIS in mid-1993. BellSouth planned to transition NMLIS to backbones based on ATM switching [9]. BellSouth also offers SMDS and frame relay as interim high-speed data networking platforms. BellSouth's long-term plans were to migrate these two services to an ATM platform.

10.2.4 NYNEX

NYNEX has been expanding its ISDN capabilities by deploying ISDN capable central office switches in additional wire centers throughout New York. NYNEX also introduced ISDN PRI in mid-1992.

To support the needs of business customers, New York Telephone announced joint plans with Northern Telecom to begin a trial of switched wideband services in June 1992. In April 1993, NYNEX introduced NYNEX enterprise services [10], a group of private lines supported by bandwidth on demand and network management capabilities. NYNEX also introduced frame relay service in 1992 using Northern Telecom's DataSPAN platform. In addition, NYNEX offered fractional T1 services at 128, 256, 384, 512 and 768 Kbps [11].

10.2.5 Pacific Bell

Pacific Bell built its telecommuting service offerings on its own experience with telecommuting. Pacific Bell developed this experience by learning from its own 1,000 employees who had been telecommuting for eight years. The services that Pacific Bell offers include the following.

- Teaching California companies the benefits of telecommuting.
- Helping telecommuters set up home offices.
- Helping businesses develop telecommuting policies [12], design pilot programs, and introduce company-wide telecommuting.
- Establishing toll-free consulting lines that allow telecommuters to seek advice on telecommuting from specially trained customer representatives on how best to set up and run a home office [13].
- Employing 250 "customer consultants" who devise ways for office workers to work at their homes, advise customers on methods to save money, and conduct meetings by telephone.

Pacific Bell has also been aggressively deploying ISDN to serve the needs of the residential and the telecommuting markets. According to industry forecasts, more than 70% of Pacific Bell's lines are planned to have ISDN access by 1995. Pacific Bell offers ISDN in the three configurations.

- An ISDN BRI service with 2B + D capabilities [14]. This resulted in cutting the installation and monthly charges significantly. The company received approval to offer the service to non-Centrex customers on a trial basis. The service was expected to be priced at $28/month.
- Centrex IS targeted to Centrex users and is offered by Pacific Bell in several configurations.
- PRI IS a primary rate service, which can support the needs of RTCs.

In addition to ISDN, in 1992, Pacific Bell introduced several new services to support the needs of the telecommuter, including switched digital service 56 (SDS 56), a circuit-switched-based digital service. Pacific Bell also introduced digital data over voice (DDOV).

To support the needs of business customers, Pacific Bell introduced a number of broadband services. These services can also meet the requirements of the RTCs. In October 1992, Pacific Bell started offering SMDS in Los Angeles, San Diego, San Francisco, and Sacramento. This service was offered following the approval of the service tariff by the Federal Communications Commission (FCC). Pacific Bell also had plans to extensively deploy frame relay and ATM-based networks extensively [3].

10.2.6 Rochester Telephone

Rochester Telephone has changed its traditional organizational structures, which classified markets into residential and business markets, and recognized the telecommuting market as a distinct market with some of the characteristics of both the residential and business markets. Rochester Telephone also established a marketing campaign with three key components: (1) market research to clearly define customer needs; (2) a strategic market plan, defining target markets, and competitive strategies; and (3) a communications plan, including traditional advertising, a direct mail campaign, and training. In addition, Rochester Telephone offered price discounts to the telecommuter market and one-stop shopping for equipment and telecommunications services.

10.2.7 U S WEST

To address the needs of the telecommuting market, U S WEST established toll-free consulting lines [12,13] that allow telecommuters to speak with specially trained customer representatives who can advise them on how best to set up and run a home office. Callers also get a full list of telco services to support the home market, such as voice mail and call waiting. In addition, U S WEST distributes a booklet, "Five secrets of a successful home office," and features point-of-sale displays highlighting its work-at-home number and services at area stores. U S WEST will also offer ISDN in response to customers' requests.

To further support the needs of the business and consumer markets, U S WEST announced plans in early 1993 to roll out broadband services (voice, video, and data). Its vision in reaching the consumer market is based on a combination of fiber and coax infrastructure. To accelerate the realization of this vision, and to expand its presence beyond its own territories, U S WEST entered into an alliance with Time Warner, the largest cable TV company. This alliance, which was created through an investment of $2.5 billion by U S WEST in Time Warner, will enable U S WEST to offer communications, entertainment, and information services to homes and businesses [15].

To further meet the needs of its business customer, U S WEST extensively deployed SMDS and frame relay and introduced a transparent LAN service (TLS) [16].

10.2.8 Southwestern Bell Telephone Company

To support the ISDN market, Southwestern Bell recently announced construction of an overlay infrastructure that will provide ISDN to all users in the Austin, Texas, metropolitan area by June 1994 [17]. To meet the growing communications needs of business customers, Southwestern Bell also announced plans to deploy SONET. They plan to first deploy this technology in areas of new customer growth and to support self-healing services. In December 1991, Southwestern Bell began testing SMDS and frame relay in preparation for deployment. Frame relay is currently available at least in 12 cities. In mid-1992, Southwestern Bell introduced circuit-switched-based videoconferencing services at 56 Kbps, 384 Kbps, and 1.5 Mbps [18].

10.3 THE INTEREXCHANGE CARRIERS

The three IXCs with the most extensive services to the consumer and business markets are AT&T, MCI, and Sprint. These IXCs have several similar product characteristics:

1. They support circuit-switched digital services. After years of questioning the reliability and availability of this technology, data communications managers are finally acknowledging these switched services. In the past year, the IXCs responded to the users' needs by expanding the range of circuit-switched services that they offer.
2. They support fast-packet services. In the last few years, AT&T, MCI, and Sprint have been actively developing broadband services.
3. They support ATM deployment.
4. They have extensively deployed ISDN.

10.3.1 AT&T

AT&T offers several circuit-switched services as part of the Accunet family, including switched 56-Kbps, switched 64-Kbps, contiguous switched 384-Kbps, and switched T1 services.[3] Telecommuters can access these circuit-switched services through PRI ISDN. Initially, AT&T's service was available as a standalone service, but it was integrated in its software-defined network (SDN) in early 1992.

AT&T introduced frame relay as part of its Interspan family of products. The Interspan FRS was made available in 1Q92 on a limited basis, and was generally made available in 2Q92. The access speeds of FRS are up to 1.5 Mbps. AT&T also developed X.25-frame relay interoperability under the Interspan umbrella. AT&T differentiated its FRS by offering users exceptional network management capabilities through an SNMP-based network management system.

10.3.2 MCI

MCI entered the telecommuting market in 1992 when it learned that an increasing number of employees of its large customers were working from their homes. MCI offers a number of services to the telecommuting market, including consulting services to help corporations with their home-based programs by customizing MCI communications systems to fit these programs and an 800 service [19,20] introduced in March 1993 and called Home Office Link. This service works as follows: (1) Callers to an 800 Home Office Link number first hear a company greeting, which is followed by a request to enter an employees' personal code on their touch-tone keypads; (2) if there is no response, MCI transfers the call to a phone mail system.

Besides specific services targeting the telecommuting market, MCI offers switched 56 Kbps and switched 64 Kbps. Both require ISDN access for switched data services. MCI also offers switched T-1 and switched T3 services.

To support the networking needs of the business market, MCI developed a cell-based switching platform that will enable it to support next-generation data technologies, including frame relay, SMDS, and ATM. In 1992, MCI introduced FRS, which was made available in 200 to 300 cities. MCI also supports SMDS and ATM, and the company recently announced tariffs for nationwide service.

10.3.3 Sprint

Sprint offers switched 56 Kbps and switched 64 Kbps, as well as ISDN. Sprint does not currently have a switched T-1 offering. Instead, it opts to supply the same capacity via N × 56/64 Kbps circuits.

---

3. This information is based on published reports; although it is believed to be correct, it has not been verified with the carriers.

To support the needs of business customers, Sprint was an early adopter of SONET technologies. Sprint was also among the early adopters of the frame relay technology. Sprint introduced the service in 1991 and made it available at all domestic POPs, in the United Kingdom, and in Japan. Sprint plans to carry SMDS over its frame relay network.

10.4 THE NEW ENTRANTS

Two groups of communications service providers are likely entrants into the telecommuting market in the next few years: the cable TV companies and the alternate access providers.

10.4.1 The Cable TV Companies

A number of cable TV companies are likely entrants in the telecommuting market for several reasons: (1) they have the cable/fiber infrastructure that interconnects their networks with tens of millions of subscribers; (2) they have the bandwidth that is required to meet the needs of the telecommuters; and (3) they have the billing systems associated with an extensive customer base.

TCI is the one of the cable TV companies that has already entered the telecommuting market. TCI refers to telecommuting as cablecommuting [21]. To offer telecommuting services, TCI entered into a partnership with Sun Microsystems Inc. and Stanford University to conduct trials in San Jose, California. TCI uses cable as a downstream, one-way link to homes with the telephone network as the return path. Telecommuters, who must be cable subscribers, are connected to the head end via a modem that is connected to their PCs or through a card that is loaded into the machine. Data is received via a standard cable channel and is transmitted as any basic or pay channel, without affecting any other cable services. On the corporate site, the company's computer is also connected to the cable system's headend via a modem. Once a corporate computer is connected to the cable's head end, access to the corporate computer is available to any telecommuter with little incremental cost to the cable system.

TCI envisions the future as the connection between cable and the public telephone network. The cable TV company is responsible for providing necessary security access.

10.4.2 The Alternate Access Providers

The alternate access providers (AAP) are the second group of potential new entrants in the telecommuting market. There are currently over 40 AAPs in the communications service industry. This number may decline significantly in the next few years as a result of industry consolidation.

The largest two AAPs are Teleport and Metropolitan Fiber Systems (MFS). These two AAPs generated about $140 million in revenues from telecommunications services in 1993. These revenues were generated by the two AAPs by offering a number of telecommunications services to large businesses in major urban areas throughout the United States. These services include private FT1, T1, and T3 services, and NMLIS.

Considering that AAPs have been focusing primarily on large businesses, AAPs are likely providers of telecommunications services to RTCs.

10.5 TELECOMMUTING BUSINESS OPPORTUNITIES FOR NETWORK SERVICE PROVIDERS

The telecommuting market represents attractive business opportunities for a number of communications services. In the short term, ISDN could represent a business opportunity to ISDN equipment and service providers for the following reasons.

- ISDN BRI addresses the needs of corporate agents and marketing and administrative telecommuters for voice communications and light to medium data communications.
- ISDN PRI addresses the needs of RTCs for high-bandwidth voice communications and data communications.
- Demand for ISDN is driven by the needs of all telecommuters for high performance and reliable communications service.
- ISDN has several competitive advantages over substitute products, including higher reliability, higher bandwidth, lower transmission cost, and a single access line.

The success of ISDN is not guaranteed yet. ISDN will be facing new challenges in the next few years. These challenges include the following.

- The emergence of other competitive alternatives, such as cable TV solutions.
- Providing telecommuters with access to VANs and the Internet.
- ISDN should be ubiquitous, or at least available in highly concentrated residential areas.
- ISDN CPE prices need to drop significantly so that they can compete effectively with prices of modems.
- Equipment distribution channels should be developed by the communications service providers and/or the CPE manufacturers so that the telecommuters can easily find and select the variety of CPE that they need to establish connections between their homes and their corporate offices.
- The ordering and provisioning of ISDN should continue to be simplified.

While ISDN could represent an important business opportunity for communications service providers in the short run, in the long run, and assuming the growth of the RTC trend, RTCs provide business opportunities for the following communications services:

- SMDS;
- Frame relay service;
- Cell relay service;
- NMLIS;
- Integrated access.

References

[1]. "ISDN Deployment: The Tortoise Makes Its Move," *Data Communications*, Sept. 1993, p. 17.

[2] "Ameritech, Bell Atlantic Maintain Rapid ISDN Pace with Intel Deals," Feb. 24, 1993, pp. 4–5.

[3] "Frame Relay, SMDS, ATM Coming Fast, Trials and Trends Reveal," Jan. 7, 1993. p. 4.

[4] "Ameritech Rolls Out The Rings," *Telephony*, March 15, pp. 9–10.

[5] "Bell Atlantic Launches Full-Scale Telecommuting Plan," *Telephone Week*, Oct. 5, 1992, pp. 4–5.

[6] "HDSL and ADSL: Steroids for the Local Loop," *Business Communications Review*.

[7] "ISDN Picks Up Steam in RBOC's Rollout Plans," *Communications News*, Feb. 1993, p. 51.

[8] "Searching for Applications, BellSouth Offers Single-Line ISDN," *Telephone Week*, Jan. 18, 1993, pp. 2–3.

[9] "Transparent LAN Ties," *Communications Week*, Feb. 22, 1993, p. 1.

[10] "NYNEX Fights to Regain Lost Customers with Innovative Service," *Telephone Week*, April 19., 1993, p. 4.

[11] "Bell's Limited Frac T-1 Supports Strands IXCs," *Intelligent Network News*, Dec. 24, 1992, p. 1.

[12] "U S WEST, Pacific Bell Eye Growth opportunities in Telecommuting," *Telephone Week*, Oct. 19, 1992, pp. 5–6.

[13] Karpinski, Richard, "Telcos Find There's 'No Place Like Home,'" *Telephony*, Oct. 19, 1992, p. 12.

[14] "Pacific Bell To Preview Single-Line ISDN At Texpo 93," *ISDN Newsletter*, March 1993, p. 18.

[15] "U S WEST Bets on Cable with Time Warner; Deal to Put Pressure on Bell Companies," *Telephone Week*, May 34, 1993, p. 8.

[16] "U S WEST Introduces LAN Interconnection Service," *Network World*, Aug. 24, 1992, p. 19.

[17] Lindstron, Annie, "California Schools In Line for ISDN," *Communications Week*, February 28, 1994, p.27.

[18] Bob Wallace. "Carriers Expand the Reach of Frame Relay Services," *Network World*, March 22, 1993, p. 4.

[19] "MCI Service Courts Work-At-Home Crowd," *Telephony*, March 15, 1993, p. 3.

[20] "MCI Targets Home Workers...And Lands '800' Contract," *Communications Week*, March 22, 1993, p. 39.

[21] "TCI Provides Stay-At-Home and Go-to-Work Service," *Metropolitan Network News*, May 25, 1992.

Implications for the Demand and Supply Sides 11

Chapter 11 focuses on reaching conclusions regarding the demand and supply sides of the telecommuting market. First, the requirements of the various types of telecommuters are compared in order to summarize the similarities and differences among these requirements. The various networking solutions that meet these diverse requirements are then compared to select the solutions that could be established as corporate standards. The chapter is concluded by proposing specific courses of action that corporate telecommuting program planners and implementers may want to consider to maximize the chances of success for their telecommuting programs.

11.1 A SUMMARY OF TELECOMMUTER NEEDS AND COMMUNICATIONS SOLUTIONS

The needs of telecommuters differ from one type of telecommuter to another. Technical telecommuters have the most extensive communications needs, including voice communications and high-bandwidth data communications as well a need to access the Internet. Marketing telecommuters need voice communications, videoconferencing, and medium-bandwidth data communications. Administrative telecommuters and home agents are the least demanding in terms of their wide area communications needs. These needs include voice communications and low-bandwidth data communications.

The communications needs of RTCs depend upon the number of participants in the RTCs and the functions that these participants perform. In any case, the communications needs of RTCs are more extensive than those of the individual telecommuters. Telecommuters currently have several networking solutions to meet their communications requirements. ISDN is currently in the best position to meet most of the requirements of the marketing telecommuters, the technical telecommuters, the home agents, and the technical telecommuters.

POTS can only adequately meet the communications needs of the least demanding telecommuters—administrative telecommuters. DOV is also an adequate solution for the administrative solution if and where available. Other solutions include hybrid frame relay/ISDN and cable TV. Consequently, a company that wants to standardize on one solution to meet the requirements of all its telecommuters should consider selecting ISDN as the standard solution.

In the long run, ADSL, ATM, and fiber/coax access solutions may represent alternatives to ISDN. These solutions, however, are not likely to be ubiquitous for many years to come.

RTCs have more networking options than individual telecommuters, including SMDS, frame relay, ATM, circuit-switched services, ISDN (PRI), and private lines.

11.2 CRITICAL SUCCESS FACTORS ON THE DEMAND SIDE

This book has identified the key networking solutions that IS managers and/or telecommuters should consider in establishing communications links with the corporate office and with other companies that employ their customers and suppliers. To maximize their chances of telecommuting programs, employers need to consider a number of other factors, which are addressed next.

1. *The goals of the telecommuting programs*: The goals of a telecommuting program have to be clearly quantified by a corporation that is contemplating the implementation of a telecommuting program. The definition of these goals should be a shaped by the functions performed by the telecommuters and their perceived value by the corporation. Examples of specific goals include improvement in productivity, employee turnover, hiring and training costs, and savings in office space. Ideally, the specific goals related to the telecommuters should be closely linked to the overall corporate goals.
2. *Careful selection of participating functions*: As described in this text, telecommuting is not appropriate for those employees whose absence from the office mean delays in urgent work and meetings. In large corporations implementing formal telecommuting programs, the selection of the appropriate functions is a joint responsibility of human resource management and the supervisors of the telecommuters.
3. *The definition of the performance objectives of the telecommuters*: Supervisors, in coordination with human resources management, need to (re)define the performance measures to reflect the results of the telecommuter's work. These results should be carefully thought through and documented to ensure that the performance evaluation is fair and to maximize the productivity of the telecommuter.

4. *Application solution selection*: As this book explained, corporations can choose among an increasing number of applications solutions, such as VANs and the Internet. Corporations will select the appropriate solutions depending upon the strength and weaknesses of each solution in relation to addressing the specific requirements of the corporation, including cost metrics.
5. *Telecommuting cost distribution*: Telecommuting costs should be distributed between employees and employers in a fair and economic fashion.
6. *Selection of the appropriate people*: Careful selection of the people who can participate in the program is mandatory for the successful implementation of a telecommuting program. Selection should be based on economic, work related, and psychological factors.
7. *Upper management support*: A formal corporate telecommuting program cannot succeed without the support of management. To ensure this success, those who are responsible for the management of a telecommuting program (e.g., human resource management) need to keep upper management informed throughout the telecommuting program development and implementation process. Results should be clearly described.

11.3 RECOMMENDATIONS TO CURRENT AND POTENTIAL TELECOMMUTERS

The success of a telecommuting program does not depend on the actions of the employers alone. The potential telecommuters also bear a responsibility towards the success of a telecommuting program. These potential telecommuters should take the decision to telecommute very seriously, considering that telecommuting is a psychological adjustment. Those employees who are considering telecommuting should be prepared to deal with the following:

- *There may be a lack of a social environment*, which is provided by the work place. Those who have a limited social life may find working at home unbearable.
- *There may be a compulsion to work all hours of the day and night*, making it difficult to distinguish between work and leisure time.
- *There may be resentment from other employees* who are not telecommuting.
- *There will be reduced visibility* in the office.
- *There may be family interruptions* and household distractions.

Those who are willing to deal with these adjustments should take several actions to ensure that telecommuting does not jeopardize their careers and/or disturb their private lives.

Those employees who have weighed the cost and benefits of working at home should consider taking the following actions to ensure the success of their experience:

- They should ensure that objectives are clearly defined and agreed upon with their supervisors.
- They should try not to telecommute on Mondays or Fridays. Instead, they should spend this time in the office to socialize with other colleagues, keep their supervisors aware of their work progress, and participate in corporate activities.

11.4 RECOMMENDATIONS TO THE NETWORK SERVICE PROVIDERS

The success of the telecommuting industry does not depend solely on telecommuters and their employers. The telecommuting industry also relies on the actions that communications service providers take to ensure its success. To maximize their revenues from the telecommuting market, these communications service providers should take the following actions:

- *Recognize the unique characteristics of the telecommuting market*, which combines some of the characteristics of the residential and the business markets.
- *Target the vertical industries* that are early adopters of telecommuting, including business and personal services, telecommunications, financial services, and manufacturing.
- *Target cities* that are expected to be early adopters of telecommuting. These are the cities that must comply with air quality standards, such as Los Angeles and Houston.
- *Offer presale and postsale services* to support customers in developing and implementing a successful telecommuting program, including maintenance, training, and consulting.
- *Establish alliances* with communications service providers to provide the telecommuters with integrated product and service offerings.
- *Offer price discounts* to stimulate the demand for these products.
- *Monitor the establishment* of neighborhood telecommuting centers.
- *Carefully develop tariffs* for telecommuters, recognizing that telecommuters are hybrid residential/business customers.

11.5 CONCLUSION

Telecommuting will likely acquire increased importance in the next few years. The principles described in this book should serve as basic guidelines for communications managers and other interested parties who will be given the responsibility of deploying the infrastructure required to support the connectivity needed to make telecommuting successful.

List of Acronyms

A

AAP	alternate access provider
ACD	automatic call distributor
ADSL	asymmetric digital subscriber line
ANI	automatic number identification
ANS	advanced network and services
ANSI	American National Standards Institute
API	application programming interface
ARPA	Advanced Research Project Agency
ASAI	adjunct/switch application interface
ATM	asynchronous transfer mode

B

BARRNET	Bay Area Regional Research Network
BISDN	broadband ISDN
BRI	basic rate interface
BSC	binary synchronous communications

C

CAD	computer aided design
CAE	computer aided engineering
CAM	computer aided manufacturing
CERFnet	California education and research federation network

CICNet	Committee on Institutional Cooperation Network
CLASS	custom local area signaling services
CLID	calling line identification
CMD	circuit mode data
CO	central office
CPE	customer premise equipment
CSU	channel service unit
CTI	computer telephone integration
CSNET	Computer Science Network

D

DEC	Digital Equipment Corporation
DLL	dynamic link library
DMT	discrete multitone transmission
DN	directory number
DNIS	dialed number identification service
DOT	Department of Transportation
DOV	data over voice
DQDB	distributed queue dual bus
DSL	digital subscriber line
DS1	digital signal 1
DS3	digital signal 3
DSU	data service unit

E

ESP	enhanced service provider

F

FDDI	fiber distributed data interface
FEP	front-end processor
FRAD	frame relay access device
FRS	frame relay service
FTP	file transfer protocol
FT1	fractional T1

H

HLI	high-speed LAN interconnection
HRM	human resource management

I

IAD	integrated access device
ID	identification
IEEE 802.6	Institute of Electrical and Electronics Engineers 802.6
IP	Internet protocol
IS	information systems
ISDN	integrated services digital network
IXC	interexchange carrier

K

Kbps	kilobits per second

L

LAN	local area network
LAPD	link access protocol for ISDN D channels
LCD	liquid crystal display
LEC	local exchange carrier

M

MAC	media access control
MAN	metropolitan area network
Mbps	megabits per second
MCI	media control interface
MIME	multipurpose Internet mail extension

N

NEARNET	New England Academic and Research Network
NFS	network file system
NI-1	National ISDN-1

NI-2	National ISDN-2
NI-3	National ISDN-3
NIUF	North American ISDN user forum
NMLIS	native mode LAN interconnection service
NSFNET	National Science Foundation Network
NSF	National Science Foundation

O

OLTP	online transaction processing
OSI	open systems interconnection

P

PAD	packet assembler/dissassembler
PC	personal computer
PIN	personal identification number
PMO	present mode of operation
POP	point of presence
PPP	point to point protocol
PRI	primary rate interface
PSTN	public switched telephone network
PVC	permanent virtual connection

Q

QoS	quality of service

R

RBOC	regional Bell operating company
RPC	remote procedure call
RTC	regional telecommuting center

S

SLIP	serial line internet protocol
SMDS	switched multimegabit digital network

SMTP	simple mail transfer protocol
SNA	system network architecture
SNI	subscriber network interface
SONET	synchronous optical network
SPI	service provider interface
SPID	service profile Identifier
SURANet	Southeastern Universities Association Network
SVC	switched virtual connection

T

TA	terminal adapter
TAPI	telephone application programming interface
TCP	transmission control protocol
TDM	time division multiplexing
TLS	transparent LAN service

U

UMA	unified messaging architecture
UDP	user datagram protocol

V

VAN	value-added network
VPU	voice processing unit
VRU	voice response unit

W

WAIS	wide area information services
WAN	wide area network
Westnet	Southwestern States Network
WOSA	Windows open services architecture

Bibliography

Green, Thomas A., "Pay Attention to Your Partners," *Telephony,* May 10, 1993, pp. 26–29.

Minoli, Daniel, "Technology Overview: Frame Relay," Datapro Communications Series: Broadband Networking, July 1993, pp. 2830–2831.

About The Authors

Osman E. Eldib

Mr. Eldib has extensive experience in a wide range of telecommunications technologies, applications, and markets. He has acquired this experience through the management and implementation of a wide range of studies with Bellcore (7 years) and Corning Glass Works (2 years). The major areas of Mr. Eldib's expertise are:

1. Segmentation and application analysis of three vertical industries: higher education, federal government, and state government;
2. Cross-industry application analysis (e.g., distance learning, CAD/CAM, electronic messaging, and geographic information systems);
3. Analysis of network management and disaster recovery requirements, and service opportunities in the financial industry;
4. Strategic business planning support of an emerging LEC broadband service;
5. Audit and comparison of the capabilities of the five leading interexchange carriers associated with the following technologies: virtual private networks, SS7, billing, network management, circuit switching, fast-packet switching, fiber optics, and microwave;
6. Strategic planning of new fiber-optic products sensors and passive components (connectors and wave division multiplexers);
7. Competitive analysis of the strategies of major alternate access providers and interexchange carriers and of the implications for the local exchange carriers.

Mr. Eldib holds an MBA degree from NYU and B.Sc. degree in engineering from the University of Alexandria, Egypt. Prior to joining Corning Glass Works, Mr. Eldib held several marketing and engineering positions in the petroleum/

chemical processing industry. He is currently working on a book about distance learning to be published by Artech House.

Daniel Minoli

Mr. Minoli has all-inclusive experience in the data communications and telecommunications fields acquired at premiere companies. His specialty is network design, which aims at deploying the most cost-effective network infrastructure for an organization. For the past several years Mr. Minoli has worked on various aspects of wide area networking (WAN) technology. In addition to fundamental work in support of the development of such technology, asynchronous transfer mode (ATM) in particular, Mr. Minoli has assisted a number of Bell operating companies in issuing and reviewing a series of RFIs and RFPs to actually install ATM-based networks. He has also done work in multimedia and imaging.

Over the years, Mr. Minoli's professional responsibilities have included fundamental research, advanced network planning, traffic engineering, network design, network implementation, system integration, disaster recovery planning, communications quality control, standards work, user training, network management, technology assessment, and communications-related software development. Network designs have included ATM networks, T1 backbone networks, channel extension, LAN interconnection, frame relay networks, cell relay networks, packet-switched networks, traditional SNA networks, voice networks, radio networks, satellite networks, and international networks.

At DVI Communications, Mr. Minoli is involved with the deployment of high-end telecommunications technologies such as ATM, multimedia, imaging, and video in corporate environments for the express purpose of securing productivity enhancements. Mr. Minoli is also involved in outsourcing, as well as telecommuting and distance-learning projects.

Prior to joining DVI Communications, Mr. Minoli worked for Bell Communications Research (Bellcore), Prudential-Bache Securities, ITT World Communications, Bell Telephone Laboratories, and Network Analysis Corporation. Mr. Minoli's work at Bellcore has been aimed at supporting the internal data communications needs of the Bell operating companies; identifying new data services that can be provided in the public network; designing large end-user networks in support of responses to RFPs; ATM/BISDN/cell relay signaling standards work in support of the ATM Forum, ANSI T1 Committee, and ITU-T (formerly CCITT); architecture planning for a public ATM network supporting video and data for a major east coast telephone company; and, ATM switch selection for a central-states RBOC to deliver fast packet, video conferencing, distance-learning, and video distribution (video dialtone) services.

Documenting his wide-ranging expertise, Mr. Minoli has published several communication and computer books that have enjoyed high circulation and critical acclaim. One book has been translated into Spanish. Mr. Minoli also published approximately two hundred technical and trade articles in numerous prestigious journals.

Mr. Minoli is a frequent speaker and session organizer at industry conferences and he is quoted regularly in the trade and in the general press. He is an adjunct associate professor at New York University's Information Technology Institute, where he has educated over 1,200 professionals over a period of ten years. He has lectured at Carniege-Mellon University's Information Networking Institute, Stevens Institute of Technology, and at the Rutgers Center for Management Development. He is on Datapro's Advisory Board for broadband networking, and over the last decade has published over 50 key technology "reports." Mr. Minoli has written several reports for Probe Research Corporation. Mr. Minoli is also a contributor editor of *Network Computing* magazine and a columnist for *Network World.* He has run a quarterly two-day seminar on T1 Communication for NYU for a number of years and more recently has delivered individually developed multisite video seminars for MCI, AT&T, Bellcore, and the RBOCs. In the past, Mr. Minoli has acted as a columnist for *Computer-World Magazine* and has been a reviewer for the *IEEE.*

Index

The Artech House Telecommunications Library

Vinton G. Cerf, Series Editor

Advanced Technology for Road Transport: IVHS and ATT, Ian Catling, editor

Advances in Computer Communications and Networking, Wesley W. Chu, editor

Advances in Computer Systems Security, Rein Turn, editor

Advances in Telecommunications Networks, William S. Lee and Derrick C. Brown

Analysis and Synthesis of Logic Systems, Daniel Mange

Asynchronous Transfer Mode Networks: Performance Issues, Raif O. Onvural

ATM Switching Systems, Thomas M. Chen and Stephen S. Liu

A Bibliography of Telecommunications and Socio-Economic Development, Heather E. Hudson

Broadband: Business Services, Technologies, and Strategic Impact, David Wright

Broadband Network Analysis and Design, Daniel Minoli

Broadband Telecommunications Technology, Byeong Lee, Minho Kang, and Jonghee Lee

Cellular Radio: Analog and Digital Systems, Asha Mehrotra

Cellular Radio Systems, D. M. Balston and R. C. V. Macario, editors

Client/Server Computing: Architecture, Applications, and Distributed Systems Management, Bruce Elbert and Bobby Martyna

Codes for Error Control and Synchronization, Djimitri Wiggert

Communications Directory, Manus Egan, editor

The Complete Guide to Buying a Telephone System, Paul Daubitz

Computer Telephone Integration, Rob Walters

The Corporate Cabling Guide, Mark W. McElroy

Corporate Networks: The Strategic Use of Telecommunications, Thomas Valovic

Current Advances in LANs, MANs, and ISDN, B. G. Kim, editor

Digital Cellular Radio, George Calhoun

Digital Hardware Testing: Transistor-Level Fault Modeling and Testing, Rochit Rajsuman, editor

Digital Signal Processing, Murat Kunt

Digital Switching Control Architectures, Giuseppe Fantauzzi

Distributed Multimedia Through Broadband Communications Services, Daniel Minoli and Robert Keinath

Disaster Recovery Planning for Telecommunications, Leo A. Wrobel

Document Imaging Systems: Technology and Applications, Nathan J. Muller

EDI Security, Control, and Audit, Albert J. Marcella and Sally Chen

Electronic Mail, Jacob Palme

Enterprise Networking: Fractional T1 to SONET, Frame Relay to BISDN, Daniel Minoli

Expert Systems Applications in Integrated Network Management, E. C. Ericson, L. T. Ericson, and D. Minoli, editors

FAX: Digital Facsimile Technology and Applications, Second Edition, Dennis Bodson, Kenneth McConnell, and Richard Schaphorst

FDDI and FDDI-II: Architecture, Protocols, and Performance, Bernhard Albert and Anura P. Jayasumana

Fiber Network Service Survivability, Tsong-Ho Wu

Fiber Optics and CATV Business Strategy, Robert K. Yates *et al.*

A Guide to Fractional T1, J. E. Trulove

A Guide to the TCP/IP Protocol Suite, Floyd Wilder

Implementing EDI, Mike Hendry

Implementing X.400 and X.500: The PP and QUIPU Systems, Steve Kille

Inbound Call Centers: Design, Implementation, and Management, Robert A. Gable

Information Superhighways: The Economics of Advanced Public Communication Networks, Bruce Egan

Integrated Broadband Networks, Amit Bhargava

Intelcom '94: The Outlook for Mediterranean Communications, Stephen McClelland, editor

International Telecommunications Management, Bruce R. Elbert

International Telecommunication Standards Organizations, Andrew Macpherson

Internetworking LANs: Operation, Design, and Management, Robert Davidson and Nathan Muller

Introduction to Error-Correcting Codes, Michael Purser

Introduction to Satellite Communication, Bruce R. Elbert

Introduction to T1/T3 Networking, Regis J. (Bud) Bates

Introduction to Telecommunication Electronics, A. Michael Noll

Introduction to Telephones and Telephone Systems, Second Edition, A. Michael Noll

Introduction to X.400, Cemil Betanov

Land-Mobile Radio System Engineering, Garry C. Hess

LAN/WAN Optimization Techniques, Harrell Van Norman

LANs to WANs: Network Management in the 1990s, Nathan J. Muller and Robert P. Davidson

Long Distance Services: A Buyer's Guide, Daniel D. Briere

Measurement of Optical Fibers and Devices, G. Cancellieri and U. Ravaioli

Meteor Burst Communication, Jacob Z. Schanker

Minimum Risk Strategy for Acquiring Communications Equipment and Services, Nathan J. Muller

Mobile Communications in the U.S. and Europe: Regulation, Technology, and Markets, Michael Paetsch

Mobile Information Systems, John Walker

Narrowband Land-Mobile Radio Networks, Jean-Paul Linnartz

Networking Strategies for Information Technology, Bruce Elbert

Numerical Analysis of Linear Networks and Systems, Hermann Kremer *et al.*

Optimization of Digital Transmission Systems, K. Trondle and Gunter Soder

Packet Switching Evolution from Narrowband to Broadband ISDN, M. Smouts

Packet Video: Modeling and Signal Processing, Naohisa Ohta

Personal Communication Systems and Technologies, John Gardiner and Barry West, editors

The PP and QUIPU Implementation of X.400 and X.500, Stephen Kille

Principles of Secure Communication Systems, Second Edition, Don J. Torrieri

Principles of Signaling for Cell Relay and Frame Relay, Daniel Minoli and George Dobrowski

Principles of Signals and Systems: Deterministic Signals, B. Picinbono

Private Telecommunication Networks, Bruce Elbert

Radio-Relay Systems, Anton A. Huurdeman

Radiodetermination Satellite Services and Standards, Martin Rothblatt

Residential Fiber Optic Networks: An Engineering and Economic Analysis, David Reed

Secure Data Networking, Michael Purser

Service Management in Computing and Telecommunications, Richard Hallows

Setting Global Telecommunication Standards: The Stakes, The Players, and The Process, Gerd Wallenstein

Smart Cards, José Manuel Otón and José Luis Zoreda

Television Technology: Fundamentals and Future Prospects, A. Michael Noll

Telecommunications Technology Handbook, Daniel Minoli

Telecommuting, Osman Eldib and Daniel Minoli

Telephone Company and Cable Television Competition, Stuart N. Brotman

Teletraffic Technologies in ATM Networks, Hiroshi Saito

Terrestrial Digital Microwave Communciations, Ferdo Ivanek, editor

Transmission Networking: SONET and the SDH, Mike Sexton and Andy Reid

Transmission Performance of Evolving Telecommunications Networks, John Gruber and Godfrey Williams

Troposcatter Radio Links, G. Roda

UNIX Internetworking, Uday O. Pabrai

Virtual Networks: A Buyer's Guide, Daniel D. Briere

Voice Processing, Second Edition, Walt Tetschner

Voice Teletraffic System Engineering, James R. Boucher

Wireless Access and the Local Telephone Network, George Calhoun

Wireless Data Networking, Nathan J. Muller

Wireless LAN Systems, A. Santamaría and F. J. López-Hernández

Writing Disaster Recovery Plans for Telecommunications Networks and LANs, Leo A. Wrobel

X Window System User's Guide, Uday O. Pabrai

For further information on these and other Artech House titles, contact:

Artech House
685 Canton Street
Norwood, MA 02062
617-769-9750
Fax: 617-769-6334
Telex: 951-659
email: artech@world.std.com

Artech House
Portland House, Stag Place
London SW1E 5XA England
+44 (0) 171-973-8077
Fax: +44 (0) 171-630-0166
Telex: 951-659
email: bookco@artech.demon.co.uk